AF356452

Micro- and Macro-Environmental Factors in Solid Cancers

Micro- and Macro-Environmental Factors in Solid Cancers

Editor

Elda Tagliabue

MDPI • Basel • Beijing • Wuhan • Barcelona • Belgrade • Manchester • Tokyo • Cluj • Tianjin

Editor
Elda Tagliabue
Head of Molecular Targeting
Unit, Department of Research,
AmadeoLab, Fondazione IRCCS
Istituto Nazionale dei Tumori
Italy

Editorial Office
MDPI
St. Alban-Anlage 66
4052 Basel, Switzerland

This is a reprint of articles from the Special Issue published online in the open access journal *Cells* (ISSN 2073-4409) (available at: https://www.mdpi.com/journal/cells/special_issues/factors_cancers).

For citation purposes, cite each article independently as indicated on the article page online and as indicated below:

LastName, A.A.; LastName, B.B.; LastName, C.C. Article Title. *Journal Name* **Year**, *Volume Number*, Page Range.

ISBN 978-3-0365-0692-0 (Hbk)
ISBN 978-3-0365-0693-7 (PDF)

Contents

About the Editor

Elda Tagliabue, Ph.D., graduated in Biological Sciences from the University of Milan in 1978, and in 1982, she obtained her Ph.D. in Biomolecular Sciences. From 1983 to 2008, she was a Staff Scientist at Fondazione IRCCS Istituto Nazionale dei Tumori of Milan (INT), and, since 2009, she has been Head of the Molecular Targeting Unit at the Research Department of INT. She is involved in both clinical and pre-clinical research addressing the role of oncoproteins and receptors for adhesion molecules in breast cancer tumorigenesis, progression and response to therapy. Concerning the study of tumor/stroma interactions, Dr. Tagliabue strongly contributed to shedding light on the significance of the extracellular matrix in breast cancer progression and so opened a discussion on the clinical relevance of stromal signatures.

Editorial

Special Issue: Micro- and Macro-Environmental Factors in Solid Cancers

Elda Tagliabue

Molecular Targeting Unit, Department of Research, Fondazione IRCCS Istituto Nazionale dei Tumori di Milano, 20133 Milan, Italy; elda.tagliabue@istitutotumori.mi.it

Academic Editor: Alexander E. Kalyuzhny
Received: 25 January 2021; Accepted: 26 January 2021; Published: 27 January 2021

Paracrine and endocrine signaling between the tumor and host have been convincingly shown to support tumor progression. As cancer development depends on stochastic mutational events, knowledge regarding the signal(s) that pass between tumor cells, the surrounding microenvironment, and host organs at distant anatomic sites is essential for improving diagnostic, prognostic, and therapeutic approaches.

This Special Issue comprises six papers, including three original articles and three reviews, covering a broad range of topics investigating the host factors that support solid tumors, highlighting the relevance of micro- and macro-environmental factors in solid cancers.

The review by Rybinska et al. [1] skillfully summarizes the current knowledge regarding tumor–adipocyte crosstalk. Recently, adipose tissue has developed from being considered an inert energy storage site to being recognized as an endocrine organ that functions in hematopoiesis, lymphopoiesis, immune function, and reproduction. As associations between obesity and some cancers have been suggested and because some tumors appear to spread to adipose-rich sites, the possibility of adipocytes sustaining the progression of tumors that develop in proximity to these cells, as in breast cancer (BC), has emerged. The analysis of adipocytes surrounding tumors has revealed a lack of lipids and the acquisition of fibroblast-like features, supporting the occurrence of intimate crosstalk between cancer cells and adipocytes. Tumor-modified adipocytes (also known as cancer-associated adipocytes (CAAs)) comprise a large population of stromal cells in BC, which are entirely committed to sustaining tumor progression and promoting resistance to treatment.

Although the precise mechanism(s) involved in the crosstalk between cancer cells and adipocytes remains to be determined, adipocytes that have previously been conditioned by tumor cells are recognized as releasing molecules that can stimulate tumor invasiveness. Recent evidence has demonstrated that CAAs feature an altered secretome compared with that of mature adipocytes, similar to obese adipocytes, which suggests that the increased tumor incidence in obese people is an effect of altered adipocytes. Thus, the targeting of adipocytes or their crosstalk with tumor cells deserves further investigation to improve tumor therapy and prevention strategies.

The micro-environmental factors that support solid tumors represent a complex network that includes both host cellular and molecular components. The extracellular matrix (ECM) surrounding solid tumors has been reported to play relevant roles in tumor development and progression. The ECM consists of a network of biochemically distinct components, including fibrous proteins, glycoproteins, proteoglycans, and polysaccharides, and the tumor ECM has been shown to differ significantly from that in normal organs. The ECM influences intratumoral signaling, transport mechanisms, and metabolism, as well as affects neoplastic cell dissemination and the development of resistance to therapy.

Palumbo et al. [2] provide a thorough review of the spectrum of interactions potentially mediated by the ECM in esophageal cancer (EC). EC is a highly lethal cancer, and the current knowledge of EC biology remains limited. Similar to most solid tumors, in EC, the interaction between tumor cells and the ECM governs various phenomena that are crucial for tumor progression. The ECM molecules and signaling pathways that have been identified as being altered in EC include lysyl oxidase (LOX) and the matrix metalloproteinases (MMPs). These proteins are fundamental for collagen turnover and therefore regulate ECM stiffness. Increased ECM stiffness promotes tumor cell growth by inducing telomerase activity, followed by telomere elongation, which sustains a limitless replication pattern that is accompanied by genetic instability. The increased production of ECM molecules has been reported to improve tumor cell adhesion, resulting in the initiation of intracellular signaling pathways, which are crucial for tumor cell survival, migration, and the activation of oncogenic signaling. Conversely, increased MMP levels can be triggered by active signaling pathways in tumor cells, such as the mitogen-activated protein kinase (MAPK) pathway.

This review also focuses on the crosstalk between EC cells and the ECM as a likely mechanism through which risk factors that are associated with EC (obesity and gastroesophageal reflux disease) promote tumor survival. Increases in leptin and interleukin-17 (IL-17) levels, which are induced by these risk factors, support higher MMP expression and activity by promoting the development of an inflammatory environment. As the cellular mechanism that underlies the development of EC has not yet been fully elucidated, the documented function of the ECM during EC progression may represent a basis for improving EC management strategies in the future.

Pancreatic ductal adenocarcinoma (PDAC) is among the tumors with a strong desmoplastic reaction, and is associated with high ECM deposition, which has been found to promote both PDAC aggressiveness and resistance to therapy.

Resovi et al. [3] investigated the therapeutic value of targeting connective tissue growth factor (CCN2/CTGF), a profibrotic matricellular protein that is highly expressed in the PDAC microenvironment and is associated with disease progression. CCN proteins are produced and secreted by many tumors and bind to various ECM molecules and cellular receptors (e.g., Notch, integrins, TrkA, and heparan sulfate proteoglycans) to modulate various activities, including replication, death, adhesion, motility, and ECM production. Given its role in PDAC progression, CCN2 has recently been proposed to serve as a therapeutic target for this aggressive tumor. The authors investigated the therapeutic value of two modified synthetic peptides that were derived from the active regions of CCN3, which is an endogenous inhibitor of CCN2. Peptide treatment in a murine orthotopic PDAC model impaired collagen deposition in the tumor microenvironment and increased chemotherapeutic activity, revealing new therapeutic perspectives for one of the most aggressive tumor types with limited treatment options.

The tumor stroma and leukocyte infiltration are widely recognized as supporting tumor progression by fueling inflammation, and the administration of non-steroidal anti-inflammatory drugs (NSAIDs) after diagnosis could increase long-term survival. Prostaglandin-endoperoxide synthase-2 (PTGS2), one of the key enzymes that mediate prostaglandin neosynthesis, is typically induced by inflammatory stimuli. PTGS2 is considered an ideal target for reducing tumor inflammation, especially in colorectal cancer (CRC) patients. Despite an ongoing, multicenter, Phase III trial to evaluate the efficacy of a PTGS2 inhibitor, celecoxib, combined with folinic acid, fluorouracil, and oxaliplatin (FOLFOX) chemotherapy, established criteria for patient selection are still lacking.

By analyzing PTGS2 levels in a series of 100 primary CRC patients using immunohistochemistry (IHC) and immunofluorescence double-staining assays, Venè et al. [4] show that cancer-associated fibroblasts (CAFs) represent a prevalent, non-tumor source of PTGS2. The in vitro analysis of primary CRC-associated CAFs revealed the powerful induction of PTGS2 expression by IL1β treatment, which is an important mediator of the inflammatory response. Intermediate levels of stromal PTGS2 were associated with better prognosis, whereas high PTGS2 levels were associated with a negative outcome compared with low- or null-PTGS2-expressing CRCs. In contrast, tumor-associated PTGS2, assessed by IHC, did not influence patient outcomes. By verifying the importance of the stroma during tumor progression, they identified an unexpected association between stromal PTGS2 levels and patient prognosis, which supports the potential role of stromal PTGS2 as a predictive marker for the response to NSAID treatment.

Cosentino et al. [5] provide evidence that BC can corrupt the surrounding microenvironment through the release of microRNAs (miRNAs) into the stroma. Specifically, miR-9, released by triple-negative BC (TNBC) cells, was found to perturb the transcriptional landscape of fibroblasts, inducing a shift toward a CAF malignant phenotype through the downregulation of the epidermal growth factor (EGF)-containing fibulin extracellular matrix protein 1 (EFEMP1). TNBC cells that were conditioned with the supernatant of normal fibroblasts that had been transfected with miR-9 or silenced for EFEMP1 became more resistant to cisplatin treatment. EFEMP1 is a structural protein that also interacts with the tissue inhibitor of metalloproteinase-3 (TIMP-3), which, in turn, inhibits the metalloproteinases MMP2 and MMP9, which are highly expressed in breast cancers and are actively involved in matrix remodeling. This study highlights the relevance of miRNAs in tumor cell communication with the host. In addition, because released miRNAs can enter the blood, they provide compelling support that tumors can corrupt the host.

The cancer cells within a tumor are now recognized as consisting of a heterogenous population, and a small subset of cancer cells, cancer stem cells (CSCs), has been identified as a reservoir of self-sustaining cells for tumor maintenance. Therefore, understanding the interactions between CSCs and the micro- and macro-environment is necessary for designing innovative treatments that are aimed at fighting the source of cancer. As immune tolerance is a key step in cancer development and progression, CSCs adopt specific strategies to escape attack by host immune cells, as reviewed by Castagnoli et al. [6]. Two primary mechanisms that are used by CSCs to prevent attack by immune cells include the downregulation of proteins that are involved in antigen presentation and the upregulation of immune checkpoint molecules, which impair the activity of immune cells.

CSCs can also contribute to the process of immunoediting by releasing cytokines that modulate the activity of tumor-infiltrating immune cells. In addition, CSCs were found to produce small molecules with immune-suppressive action, including prostaglandin E2, which induces a shift from a Th-1 to a Th-2 immune response. The immune checkpoint molecules that are expressed by CSCs not only inhibit immune system activation but also sustain tumor stemness by stimulating the expression of molecules that are associated with CSC-specific signaling. Soluble inflammatory cytokines that are released into the tumor microenvironment by tumor-infiltrating immune cells can also modulate CSC activities. Thus, the dynamic crosstalk that occurs between CSCs and the immune tumor microenvironment represents a key player in allowing CSCs to escape host immune recognition, sustaining tumor maintenance and expansion. These results support several ongoing trials that are testing the efficacy of combining therapies against CSCs with existing therapies that use immune-modulating agents that promote an effective antitumor immune response.

Recognizing that many aspects of tumor biology can be explained by the dynamic crosstalk that occurs between the tumor and the host, we hope that this Special Issue will be of interest, particularly to researchers who are focused on tumor–host interactions. The findings from these studies are expected to provide a better understanding of how local and systemic environments contribute to cancer progression and to serve as the basis for novel diagnostic and therapeutic approaches.

Funding: The author received no specific funding for this editorial.

Conflicts of Interest: The author declares no conflict of interest.

References

1. Rybinska, I.; Agresti, R.; Trapani, A.; Tagliabue, E.; Triulzi, T. Adipocytes in Breast Cancer, the Thick and the Thin. *Cells* **2020**, *9*, 560. [CrossRef] [PubMed]
2. Palumbo, A., Jr.; Meireles Da Costa, N.; Pontes, B.; Leite de Oliveira, F.; Lohan Codeco, M.; Ribeiro Pinto, L.F.; Nasciutti, L.E. Esophageal Cancer Development: Crucial Clues Arising from the Extracellular Matrix. *Cells* **2020**, *9*, 455. [CrossRef] [PubMed]
3. Resovi, A.; Borsotti, P.; Ceruti, T.; Passoni, A.; Zucchetti, M.; Berndt, A.; Riser, B.L.; Taraboletti, G.; Belotti, D. CCN-Based Therapeutic Peptides Modify Pancreatic Ductal Adenocarcinoma Microenvironment and Decrease Tumor Growth in Combination with Chemotherapy. *Cells* **2020**, *9*, 952. [CrossRef] [PubMed]
4. Vene, R.; Costa, D.; Augugliaro, R.; Carlone, S.; Scabini, S.; Casoni, P.G.; Boggio, M.; Zupo, S.; Grillo, F.; Mastracci, L.; et al. Evaluation of Glycosylated PTGS2 in Colorectal Cancer for NSAIDS-Based Adjuvant Therapy. *Cells* **2020**, *9*, 683. [CrossRef] [PubMed]
5. Cosentino, G.; Romero-Cordoba, S.; Plantamura, I.; Cataldo, A.; Iorio, M.V. miR-9-Mediated Inhibition of EFEMP1 Contributes to the Acquisition of Pro-Tumoral Properties in Normal Fibroblasts. *Cells* **2020**, *9*, 2143. [CrossRef] [PubMed]
6. Castagnoli, L.; De Santis, F.; Volpari, T.; Vernieri, C.; Tagliabue, E.; Di Nicola, M.; Pupa, S.M. Cancer Stem Cells: Devil or Savior-Looking behind the Scenes of Immunotherapy Failure. *Cells* **2020**, *9*, 555. [CrossRef] [PubMed]

Publisher's Note: MDPI stays neutral with regard to jurisdictional claims in published maps and institutional affiliations.

Review

Adipocytes in Breast Cancer, the Thick and the Thin

Ilona Rybinska [1], Roberto Agresti [2], Anna Trapani [2], Elda Tagliabue [1] and Tiziana Triulzi [1,*

[1] Molecular Targeting Unit, Department of Research, Fondazione IRCCS Istituto Nazionale dei Tumori, Milan 20133, Italy; ilona.rybinska@istitutotumori.mi.it (I.R.); elda.tagliabue@istitutotumori.mi.it (E.T.)

[2] Division of Surgical Oncology, Breast Unit, Fondazione IRCCS Istituto Nazionale dei Tumori, Milan 20133, Italy; roberto.agresti@istitutotumori.mi.it (R.A.); anna.trapani@istitutotumori.mi.it (A.T.)

* Correspondence: tiziana.triulzi@istitutotumori.mi.it; Tel.: +39-022-390-5121

Received: 30 January 2020; Accepted: 26 February 2020; Published: 27 February 2020

Abstract: It is well established that breast cancer development and progression depend not only on tumor-cell intrinsic factors but also on its microenvironment and on the host characteristics. There is growing evidence that adipocytes play a role in breast cancer progression. This is supported by: (i) epidemiological studies reporting the association of obesity with a higher cancer risk and poor prognosis, (ii) recent studies demonstrating the existence of a cross-talk between breast cancer cells and adipocytes locally in the breast that leads to acquisition of an aggressive tumor phenotype, and (iii) evidence showing that cancer cachexia applies also to fat tissue and shares similarities with stromal-carcinoma metabolic synergy. This review summarizes the current knowledge on the epidemiological link between obesity and breast cancer and outlines the results of the tumor-adipocyte crosstalk. We also focus on systemic changes in body fat in patients with cachexia developed in the course of cancer. Moreover, we discuss and compare adipocyte alterations in the three pathological conditions and the mechanisms through which breast cancer progression is induced.

Keywords: adipocytes; breast cancer; obesity; adipokines; cancer associated adipocytes (CAAs); cachexia

---

1. Introduction

In cancer, the tumor surrounding cells have altered biological properties compared with the normal state. We know today that there is a large spectrum of reciprocal interactions between a tumor and its stroma that significantly influence tumor biology and support tumor progression. Given the close juxtaposition of adipocytes and breast cancer (BC) cells, it is somehow surprising that knowledge-intensive research about the role of adipocytes in tumor initiation, growth, and metastasis is a relatively new area of investigation [1]. The previous lack of attention can be partially explained by the fact that adipose tissue (AT) used to be regarded as a rather inert site principally storing energy in the form of lipids and cushioning the body [2]. It has taken on a new significance, particularly since the discovery of leptin in 1994 [3]. Thereafter, the list of adipocyte-derived factors has been increasing at an extraordinary pace, and the traditional role of AT has progressed to a fully functioning endocrine organ, acting on local and systemic levels and modulating feeding behavior, total body energy expenditure, andsuch fundamental processes as hematopoiesis, lymphopoiesis, immune function, and reproduction [4,5]. A great importance attributed to the role of adipose tissue in cancer is sustained by two main observations: (i) epidemiologic studies have demonstrated an association between obesity and some cancers [4], and (ii) many tumors metastasize to adipose rich niches such as abdomen or bone marrow [6,7]. In addition, adipose mass in these sites is increased with obesity and aging and in turn may explain cancer progression in obese and/or elderly patients [7]. Worth mentioning, cancer cachexia, which is due to severe metabolic dysregulation in fat tissue, is observed in more than 50% of cancer patients, including those with BC [8]. Importantly, adipose tissue wasting observed in cancer cachexia shares some similarities with stromal-carcinoma metabolic synergy [9]. Finally, it is obvious that, in organs such as the breast, early local tumor invasion results in immediate cancer cell proximity to adipocytes. The large number of recently published studies supported the presence of the heterotypic interaction between epithelial cells and adipocytes at the invasive tumor front and, together with different variants of the two-compartmental in vitro models, provided more details about the mechanisms of this vicious interaction. The aim of this article is to summarize current knowledge on the role of AT and adipocytes in BC progression, focusing on mechanisms and implications of obesity, closeness between tumor cells and adipocytes, and cachexia.

2. Adipogenesis and the Adipose Organ

2.1. The Adipogenesis Process

Adipocytes derive from multipotent mesenchymal stem cells that also have the potential to differentiate in myoblasts, osteoblasts, and chondroblasts [10]. Molecular characteristics, based mainly on in vitro studies, divide adipogenesis into two stages. The first step, named determination, defines cell fate. This stage requires the commitment of a pluripotent stem cell to the adipocyte lineage. Entrance into this stage of differentiation was once considered non-reversible, but now we know that this process is multidirectional, and adipocytes under certain stimuli can return to a phenotypically stem-like precursor state. Cell growth arrest initiates the second phase of adipocyte differentiation, during which preadipocytes begin to resemble characteristics of the mature fat cells in terms of expanded lipid metabolism and the secretion of adipocyte-specific proteins. Adipogenesis in vitro is based on a cocktail of inducing agents containing 3-isobutyl-1-methylxantine (IBMX) (stimulating the cAMP-dependent protein kinase pathway), dexamethasone (stimulating the glucocorticoid receptor pathway), and insulin (serving as permissive differentiation signal to preadipocytes) required to induce differentiation of preadipocytes into fat cells [10]. Adipogenesis in its complexity is a highly regulated process, at the head of which stand the nuclear receptor-peroxisome proliferator-activated receptor-γ (PPARγ) as well as transcription co-activators CCAAT/enhancer-binding protein α and β (C/EBP α and C/EBP β) [11]. Enhanced lipid accumulation and subsequent expression of the adipocyte fatty-acid binding protein (*FABP4*) and the insulin-responsive glucose transporter type 4 (*GLUT4*) are characteristic of an early stage of differentiation [10]. Mature adipocytes additionally express adiponectin (*ADIPOQ*), leptin (*LEP*), the adipose triglyceride lipase (*ATGL*) and lipoprotein lipase (*LPL*), as well as high levels of perilipin 1 (*PLIN1*), a lipid droplet coating protein.

2.2. The Adipose Tissue

Classification, based mainly on cellular mitochondrial content, divides adipocytes in white, beige, and brown types. White adipocytes are dominant in the body and contain a large, unilocular lipid droplet and are specialized for storage of neutral lipids. Due to their unique expression of a mitochondrial membrane protein called uncoupling protein 1 (UCP1), brown adipocytes can generate endogenous heat in a process called thermogenesis. Classification of white/brown adipocytes does not fully describe their diversity as, even among white adipocytes, cells from different locations have distinct molecular and physiological properties.

Based on the anatomical distribution, AT may be divided into subcutaneous, visceral, intramuscular, bone marrow, and dermis subtypes [12]. There are significant gender-related differences in the expansion of the AT mass, but, in general, human subcutaneous AT comprises ~80% of total body fat and is located primarily in the gluteal and the femoral depots. Additionally, in women, the breast fat pad is a relevant contributor to total subcutaneous fat. On the other hand, the visceral fat portion includes omental, mesenteric, and epiploic adipose tissue as well as gonadal, epicardial, and retroperitoneal fat pads. It surrounds vital organs and makes up approximately 5% to 20% of total body fat in normal weight individuals. Although adipocytes are the major component of AT, it is composed also by preadipocytes and a stromal-vascular fraction containing endothelial cells (10–20%), pericytes (3–5%), stem and progenitor cells (0.1%), as well as a rich collection of innate and adaptive immune cells [macrophages, dendritic cells, mast cells, eosinophils, neutrophils, and lymphocytes (5–45%)] [4]. Curiously, every fat depot is really unique. More to the point, differences concern preadipocyte populations, signals governing adipogenesis, metabolic functions (secretion of adipokines and inflammatory cytokines), lipolysis rates and thermogenic potential, degree of innervation, and vascularization [10]. This is in line with their differential involvement in metabolic-related complications. Worthy of consideration, visceral fat is strongly associated with metabolic disease risk, whereas subcutaneous adiposity is comparatively benign [13].

2.3. Mammary Adipocytes

Physiological intimate bidirectional interaction of mammary adipocytes with adjacent epithelium leading to an extreme plastic phenomenon occurring in adult tissue is a hallmark of mammary adipocytes. Indeed, fat cells are required for proper mammary duct development during puberty as well as for the maintenance of ductal architecture in the adult mammary gland [14]. Moreover, during pregnancy, they modify their phenotype, making a place for alveolar cells able to synthesize various milk constituents [15]. The fate of adipocytes during this process remains hotly discussed. In particular, whether fat cells die off or transdifferentiate into another cell type, as suggested by Morroni et al. [16] is unknown. Maybe they remain constantly present as smaller, slimmed cells, giving the impression of adipocytes disappearance [16,17]. Considering the above, the post-lactation adipogenesis seems to be a process not involving proliferative events but rather based on lipid trafficking and cytoplasm refilling with components derived from regressing milk-producing cells [18]. Nowadays, we know such particular dialog persists also in pathological conditions, such as cancer.

Tumor cells exert significant effects on adjacent adipocytes. The histological picture of tumor-surrounding adipocytes unraveled that they become fewer, lose lipids, and acquire fibroblast-like features [19]. Several in vitro studies using two compartmental models (cancer cells/adipocytes) supported that the phenotype of adipocytes observed at the invasive tumor front was the result of direct influence of tumor cells [20]. The intimate cross-talk between cancer cells and adipocytes resulted in the initiation of the process of adipocytes dedifferentiation in the sense of a reduction of adipocytes terminal differentiation with a reduction in the expression of differentiation markers such as PPARγ and C/EBPα as well as their downstream genes such as *FABP4*, *ADIPOQ*, and hormone sensitive lipase (*HSL*) [19,21]. Tumor modified adipocytes (named cancer associated adipocytes, CAAs, Table 1) may be an intermediate form of arising cancer-associated fibroblasts (CAFs), which constitute the non-homogenous population comprising the majority of stromal cells in BCs [22]. Adipocyte-derived fibroblasts exhibited increased expression of the CAF marker fibroblast-specific protein-1 (FSP-1) but not alpha smooth muscle actin (a-SMA) [22].

Table 1. Comparison of adipocyte features in obesity, cachexia and after interaction with tumor cells.

Characteristic	Obesity	CAAs		Cachexia
		In Vivo	In Vitro	
Size/TG stores	↑ A size [23]	↓ A size [19] and lipid droplet size [24]	↓ lipid droplets size and number [19,24]	↓ A size and TG content [25]
Adipogenesis regulators	↓ C/EBPα and PPARγ [26]	-	↓ PPARγ and C/EBPα [19,24]	↓ PPARγ and C/EBPα [25,27]
UCP1 (browning)	↑ in brown AT [28] ↓ in white AT [29]	↑ [30]	-	↑ [31]
Glucose transport	↓ *GLUT4* [28]	-	↓ *GLUT4* and *IRS1* [24]	↓ *GLUT4* [32]
Lipogenesis	↑ in white AT [33]	↓ (↓*HOXC8, HOXC9, FABP4,* and *HSL*) [30]	↓ (↓ *FABP4, HSL, ATGL, CIDEA,* and *FASN*) [24]	↓ (↓ SREBP-1c) [32]
Lipolytic activity (*ATGL/HSL*)	↑ in subc. A [34]	↓ *HSL* [30]	↓ *HSL* [19]	↑ *HSL* [35]

A: adipocytes; AT: adipose tissue; CAAs: cancer associated adipocytes; subc: subcutaneous; TG: triglycerides.

Although some candidate molecules secreted by tumor cells such as tumor necrosis factor alfa (TNF-α) [36], Wnt3a [22], Wnt5a [37] and stromelysin-3 (MMP11) [38] have been proposed to dedifferentiate mature adipocytes, the precise mechanisms that could be involved in tumor-driven adipocyte dedifferentiation and lipid loss remain to be discovered.

3. Epidemiological/Clinical Association between Obesity and BC

According to the World Health Organization (WHO) and the National Institute of Health (NIH), overweight and obesity are clinically present when the body mass index [BMI, defined as weight (kg)/ height (m^2)] is greater than 25 or 30 kg/m^2, respectively [39]. Almost two billion adults and more than 500 million people are respectively defined as overweight and obese in the world, and these rates will increase in the future [40,41].

BC is the most frequent female type of cancer and a leading cause of cancer-related mortality worldwide [42], and it is a highly heterogeneous disease with a wide range of hysto-pathological, biomolecular patterns, and clinical behaviors that associate with different prognosis [43]. Leaving aside genetic predispositions, such as BRCA 1–2 mutations, or reproductive factors, as BC causes, tumor pathogenesis is a multifactorial process in which metabolic consequences and related interactions of an unhealthy lifestyle are epidemiologically and clinically widely studied. Surely, it is considered interesting and challenging that unbalanced diet, unsatisfactory physical activity, and high alcohol consumption contributing to determine a high BMI may be modifiable risk factors, as shown in the European Prospective Investigation into Cancer and Nutrition (EPIC) Italy study on over 15,000 post-menopausal women [44].

Two of the leading questions in this area of investigation are if there is a linear relation between increasing BMI and BC onset and what subtypes of BC are more influenced by obesity. Epidemiologically, obesity is a risk factor for many cancers [45], and it is particularly associated with BC in post-menopausal women. In a prospective cohort study within the Nurses' Health Study, more than 87,000 women were followed up, recording their weight change during a long-observed period of life and showing that weight gain since menopause significantly increases the risk of BC, particularly in obese women [46]. Other convincing evidence that body fatness and weight gain may be directly and progressively related to post-menopausal BC has been described in the larger European EPIC study on almost 250,000 post-menopausal women in which, conversely, healthy behaviors reduced the risk of BC [47]. Furthermore, evaluating in a meta-analysis the relationship of adult weight gain with subgroups of BC, Vrieling at al. showed in obese patients a significantly increased risk of post-menopausal estrogen receptor (ER)+BC [summarized risk estimate (RE) = 2.33; 95% confidential interval (CI) 2.05–2.60] [48]. This association between BMI and ER+ BC was also demonstrated by an analysis of pooled tumor markers and epidemiological risk factors in more than 35,000 invasive BC patients from 34 studies participating in the Breast Cancer Association Consortium [49].

In pre-menopausal women, studies examining the association between diet, BMI, and BC showed inconsistent results with major complexity. Suzuky et al. associated a high BMI with a 20% lower risk for ER+ BC in pre-menopausal women (95% CI = −30% to −8%), confirming an 82% higher risk in post-menopausal women (95% CI = 55–114%) [50]. The same authors showed that each five unit increase in BMI was associated with a 33% increased risk among post-menopausal women (95% CI = 20–48%) and a 10% decreased risk in pre-menopausal women (95% CI = −18% to −1%). In this paper, no association was observed for ER-negative BC, but further meta-analysis showed a significantly increased risk of triple-negative BC (TNBC) in pre-menopausal women with high BMI [51].

Some meta-analyses analyzing large numbers of patients showed that obesity is also positively related to BC recurrence and mortality. Protani et al., evaluating 43 historical studies (which principally defined obesity using BMI more than waist-to-hip ratio) and comparing obese versus non-obese patients, showed a statistically significant higher risk for overall survival (HR = 1.33) and for BC specific survival (HR = 1.33; 95% CI = 1.21–1.47) in obese patients [52]. In interest, differences were seen in pre-menopausal (HR = 1.47) and in post-menopausal (HR = 1.22), even if they were not statistically significant. A more recent meta-analysis on 82 studies evaluating more than 210,000 patients showed that BMI was significantly associated with BC mortality; for BMI calculated before the diagnosis, the summary relative risks of total mortality and of BC mortality were 1.41 (95% CI = 1.29–1.43) and 1.35 (95% CI = 1.24–1.47) for obese compared with normal weight patients, with a difference for pre-menopausal (1.75; 95% CI = 1.26–2.41) and for post-menopausal (1.34; 95% CI = 1.18–1.53) BC patients [53]. Conversely, in a further systematic review and meta-analysis, weight gain (≥ 5% of body weight, measured at least one year post-primary BC diagnosis) versus maintaining body weight during the follow-up in BC patients seemed to be associated with higher all-cause mortality but not with hazard of BC specific mortality [54]. Furthermore, in a study on 15,000 patients, overweight and obesity have been significantly associated with higher probability to develop controlateral BC (adjusted hazard ratio: 1.50; 95% CI = 1.21–1.86) compared to normal weight women after 10 years of follow-up [55].

Obviously, obese patients may have delayed diagnosis with more advanced disease at diagnosis compared with patients with normal BMI. This was shown by a study in Denmark on over 18,000 patients in which obese patients were older and had significantly more advanced BC at diagnosis than normal weight patients. However, after data adjustment for disease characteristics, obesity remained an independent prognostic factor for distant metastases and for death in BC [56]. Similarly, a French study on over than 14,000 BC patients showed that, even if obese patients presented more advanced tumors at diagnosis, multivariate analysis identified a relevant independent effect for obesity on BC recurrence [57].

4. Evidence of the Pro-Tumorigenic Effect of the Adipocyte-Tumor Cell Crosstalk

Several in vitro studies have demonstrated that AT supports and promotes tumor growth. Moreover, the degree of tumor infiltration into the adjacent fat AT serves as histological criterion reflecting the aggressiveness of the tumor and is indicative of poor prognosis [58,59]. A special role in this matter can be attributed to mature AT fraction. It was evidenced that mature adipocytes being in direct contact with BC cells, but not stromal AT preadipocytes, increased tumor growth of ER+ BC cell lines. Importantly, it is likely that mature adipocytes, during coculture with BCs for a relatively long time (7 days), acquire tumor promoting characteristics and become CAAs. This would be in line with a study carried out by Dirat et al. [19], which clearly showed that cross-talk between the two cell populations is necessary to observe the pro-invasive effect. In detail, only CAAs previously conditioned by tumor cells, not "naive" adipocytes, release molecules stimulating tumor invasiveness. Although a growth-promoting effect has been largely reported in several models, e.g., ovarian, prostate, colon, and melanoma cancers [21], the interaction between adipocytes and BC cells is very complex and often ambiguous (Table 2). For example, some BC cell lines cocultured with adipocytes exhibit increased proliferation but others did not. It can suggest that the effect depends on the tumor subtype. In contrast to proliferation, the adipocyte effect on cancer migration/invasion seems to be clearer and more repeatable in various studies. This in line with experimental studies that have shown that cocultures of adipocytes with BC cells increased cancer cell migration [60] and invasion [19]. Of note, the list of studies clearly shows that the aggressive-promoting effect of adipocytes on cancer cells is not conditioned by the direct contact, suggesting that it is caused by soluble molecules. The cancer-adipocyte crosstalk is of great importance in the modulation of tumor behavior and induction of partial epithelial mesenchymal transition (EMT) [60,61]. It was also shown that some BC cell subtypes, when cocultured with mature adipocytes, displayed a downregulation of the epithelial marker E-cadherin with a simultaneous morphological change (spindle shape) [60]. In contrast, preadipocytes were shown to induce expression of E-cadherin in BCs cell lines. In the same study, however, adipocytes did not affect E-cadherin expression in cocultured BC cells.

Table 2. Pro-tumorigenic effect of the adipocyte-tumor cell crosstalk.

Feature	Effect	Adipocyte Comments	Ref
Proliferation	↑ in ER+ (MCF-7, ZR75-1, T47-D) cells by mature A = in ER+ cells by preadipocytes = in ER- (MMT 060562) cells by mature A	Primary mature subcutaneous rat A and preadipocytes 7 days of direct contact in 3D collagen gel	[61]
	↑ in SUM159PT by mature A = in ZR 75.1, 67NR, 4T1 by mature A	3T3-F442A cell line 3 days coculture in transwell	[19]
	↑ in BC cell lines (MCF7, MDA-MB-231, N-453, N-435S, -468) by mature A	Mature 3T3-L1 2 days coculture in transwell	[60]
Tumor growth	EO771 cells injected in brown AT > white AT > subcutaneous AT	Dorsal subcutaneous fat, inguinal white AT, interscapular brown AT of C57BL/6 mice	[62]
	↑ SKOV3ip1 ovarian cancer cells injected with A vs. tumor cells alone	Human primary omental A Nude mice	[21]
EMT/ stemness	↑ E-CAD expression in T47D and MCF-7 cells by preadipocytes = E-CAD in ZR75-1 and MMT 060562 cells by preadipocytes = E-CAD expression in T47D, MCF-7, ZR75-1 and MMT 060562 cells by mature A	Primary mature subcutaneous rat A and preadipocytes 7 days of direct contact in 3D collagen gel	[61]
	↑ change in morphology in MCF-7, MDA-MB-435S, -231 cells by mature A = in MDA-MB-453, -468 cells ↓ E-CAD in MCF7 cells by mature A ↑ Vim in MCF-7, MDA-MB-231 and-435S cells by mature A	3T3-L1 A 2 days coculture in transwell	[60]
Migration/ Invasion	↑ invasion of ZR 75.1, 67NR, SUM159PT, 4T1 cells by mature A ↑ in BC cell lines by CM from CAAs = in BC cell lines by CM from "naive" A ↑ number of metastases in BALB/c mice injected with 4T1 cells previously cultivated with A	3T3-F442A cell line Indirect coculture/CM	[19]
	↑ MCF7, MDA-MB-231, -468, -453, -435S) by mature A ↑ TN (MDA-MB-231, -468, -435S) > ER+ MCF7, HER2+ MDA-MB-453	3T3-L1 A 2 days coculture in transwell	[60]
Metabolic reprogramming	↑ accumulation of lipids in T47D and MDA-MB-231 in coculture with A	Human primary omental A	[21]

A: adipocytes; AT: adipose tissue; CAA: Cancer associated adipocytes; CM: conditioned medium; EMT: epithelial mesenchymal transition; ER: estrogen receptor.

Numerous studies demonstrated that adipocyte lipolysis stimulated by cancer cells is at the very heart of the synergy between cancer cells and adipocytes. Lipolysis is defined as the hydrolytic cleavage of ester bonds in triglycerides (TGs), resulting in the generation of fatty acids (FAs) and glycerol [63]. BC cells, when cocultured with adipocytes, accumulate lipids [21]. Using the adipocyte-ovarian cancer cell coculture model, the direct transfer of lipids from adipocytes to ovarian cancer cells was proved. Moreover, it was shown, perhaps by induction of FA β-oxidation (FAO) in cancer cells, to promote tumor growth, suggesting that adipocytes act also as an energy source for cancer cells [21].

5. Mechanisms Behind the Pro-Tumorigenic Effect of Adipocytes

The pathological expansion of altered white AT in obesity leads to an abundant production of several biologically active factors, including inflammatory cytokines, hormones, adipokines, and lipid metabolites that can be considered mediators between obesity and cancer. Recent evidence demonstrated that CAAs have an altered secretome compared to mature adipocytes. CAAs share some common features with obese adipocytes, e.g., they secrete significantly higher levels of motility factors such as CCL2, CCL5, autotoxin (ATX), as well as proinflammatory cytokines such as IL-1β, IL-6, TNF-α, VEGF, and leptin. Moreover, cachectic adipocytes give an essential contribution to cachexia (Table 3).

Table 3. Molecules at the basis of the pro-tumorigenic effect of adipocytes in obesity, cachexia, and after interaction with BC.

Molecules	Obesity	CAAs		Cachexia
		In Vivo	In Vitro	
TNF-α	↑ [64]	↑ [30]	↑ [19]	↑ [65,66]
IL-6	↑ [67,68]	↑ [19]	↑ [19]	↑ [65,66]
IL-1β	↑ [69]	-	↑ [19]	↑ [66]
CCL2	↑ [70]	-	↑ [71]	↑ [66,72]
CCL5	↑ [73]	↑ [74]	-	-
Aromatase	↑ [75]	↑ [76]	-	-
Leptin	↑ [77,78]	↑ [30]	↓ [24]	↓ [79] *
Adiponectin	↓ [80] *	↓ [19,30]	↓ [19,24]	↑ [81] *
Resistin	↑ [67] *	↑ [20]	↓ [19,24]	-
Autotaxin	↑ [82–84]	↑ [85]	-	-
HGF	↑ [86]	↑ [20]	-	-
IGF-1	↑ [87,88]	-	-	-
PAI-1	↑ [89,90]	↑ [20]	↑ [19]	-
COL6A3	↑ [91]	↑ [20]	-	-
MMP11	-	↑ [20]	↑ [19,24]	-
MMPs	↑ [92]	↑ [20]	↑ (MMP9, [24])	↑ (MMP2 and MMP9, [93])
Collagen	↑ (fibrosis, [94])	-	↑ (COL I, [37])	↑ (COL I, III, VI, [27,65])

* Data derived from analyses in plasma/serum and not in AT or adipocytes.

5.1. Inflammation

It is well recognized that chronic inflammation with high circulating levels of C-reactive protein represents a pathophysiological condition that bridges obesity and cancer [95]. However, local inflammation driven by altered adipocyte also has a role in obesity. The histological biomarker of this local inflammation is represented by the presence of the so-called "crown-like structures" (CLSs), derived from dead/dying adipocytes surrounded by macrophages. Macrophages are recruited in the obesity-associated AT due to excessive accumulation of fat and production of chemokines such as CCL2 and CCL5 [70,73] and are switched towards an M1 proinflammatory state [96]. A paracrine loop involving free fatty acids (FFAs) and tumor necrosis factor-alpha (TNF-α) between adipocytes and macrophages establishes a vicious cycle that leads to high secretion of cocktail of proinflammatory mediators such as prostaglandin E2 (PGE2), TNF-α [64], IL-1β [69], and IL-6 exacerbating inflammatory status in the AT. This phenomenon has been described mainly in the visceral fat of obese women, but it occurs also in the breast AT and has a role in BC progression. Indeed, CLSs are enriched in BCs of obese patients and have negative impact on disease recurrence and survival [97]. Moreover, inflammatory factors (e.g., IL-6, IL-1β, TNF-α) are found at high levels in obese patients and are associated with poor outcome in BC patients [98]. The presence of CLSs is not limited to obese patients, but it also occurs in BCs from normal weight women [99]. Tumor trace released on adipocytes provides important modifications of fat cell secretome, which, generally speaking, becomes highly inflammatory. Indeed, CAAs secrete significant amounts of pro-inflammatory mediators such as CCL2 and IL-1β, leading to the accumulation of macrophages forming CLSs. Notably, in normal weight BCs, CLS are associated with reduced survival, supporting a role for local AT inflammation in BC progression [100,101].

IL-6 and TNF-α

In normal AT, adipocytes are not the major source of IL-6, however, under pathological conditions such as obesity and cancer, the levels of IL-6 secreted from adipocytes increase significantly [67,68]. Changes in adipocyte secretory profile (e.g., TNF-α, IL-6, and IL-1β) were also observed after coculture with BC cells or in isolated adipocytes from BCs [20]. Activated adipocytes assume an inflammatory phenotype (CAAs) and release more IL-6 [1,19]. Function of IL-6 in breast and other solid tumors progression and treatment was shown to be associated with the development of stem cell phenotype, angiogenesis, cachexia, and resistance to therapy [102]. One-third of the total circulating IL-6 is expressed predominantly by adipocytes. IL-6 circulating levels increased in human obese subjects and correlated with adiposity [67,68]. In BC patients, the extent of the increase of IL-6 in serum was correlated with poor disease outcome and reduced prognosis [103].

TNF-α is an important inflammatory factor in the tumor microenvironment that is generated by tumor and stromal cells. Moreover, BC cells in coculture with adipocytes stimulated their TNF-α production [19]. While, in the serum of healthy women, TNF-α is generally not detected, clinical studies have reported high levels of this cytokine in patients with BC [104]. It was evidenced that TNF-α may play an important role in BC development and proliferation, chemoresistance, angiogenesis, invasion, and metastasis [105]. Through upregulation of IL-6 and aromatase expression, TNF-α participates with the comprehensive regulation of estrogen synthesis [106]. In addition, TNF-α as a lipolytic agent was shown by activation MEK, ERK, and elevation of intracellular cAMP to enhance lipolysis in human adipocytes [107].

5.2. Estrogens

More than 75% of BCs express the ER, which means the vast majority of BCs grow and progress in response to estrogens. Of note, estradiol (E2), through the alteration of the tumor microenvironment, was shown to increase the growth of ER-negative BCs [108]. Estrogen biosynthesis is catalyzed by the enzyme aromatase. Most of the estrogen in pre-menopausal women is synthesized by the ovaries, while, after menopause, the adipose organ becomes the predominant source of aromatase expression and estrogen production [109]. Given the close proximity of fat tissue to epithelial cells in mammary gland tissue, aromatase derived from breast adipose stromal cells may have a substantially higher impact on breast carcinogenesis than aromatase expressed in other parts of the body. Indeed, estrogen levels in BCs are as much as 10 times greater than in the circulation of post-menopausal women [110]. It is therefore likely that locally produced estrogens, as a consequence of breast AT inflammation, may represent a key driver of post-menopausal BCs in obesity. Accordingly, aromatase activity is higher in AT of obese than normal weight women [75]. Even if direct evidence is still lacking, considering that aromatase expression in AT is a marker of preadipocytes rather than mature adipocytes and that stimulation of PPARγ leads to a reduction in aromatase expression [111], it is likely that dedifferentiated CAAs are relevant sources of aromatase within the tumor microenvironment. In support, aromatase activity was found higher in the breast quadrant containing the tumor than in the opposite one [76]. It is interesting that factors inhibiting adipocyte differentiation, such as IL-6 and TNF-α, at the same time stimulate aromatase expression in AT.

5.3. Adipokines

5.3.1. Adiponectin and Leptin

Adiponectin is the most abundant protein secreted by AT with wide range influence on several tissues and organs, performing insulin-sensitizing, anti-inflammatory, antiatherogenic, proapoptotic, and antiproliferative actions [112]. Unlike most of the other AT-derived proteins, serum adiponectin is reduced in obesity [80]. It is reduced also in CAAs in vitro and at the invasive front compared with normal mammary AT [19,30]. Adiponectin released by adipocytes in the BC microenvironment has a protective effect against tumorigenesis mainly through intracellular mechanisms initiated via its receptors (AdipoR1 and AdipoR2). It has been extensively described that adiponectin attenuates growth and invasion of BC cells and induces apoptosis by activation of AMPK and PI3K/AKT and ERK1/2 inhibition [113] rather than inducing autophagic cell death [114]. While consistent results on its anti-tumor activity have been obtained in ER-negative BC cells, controversial have been reported in ER+ BC cells [115].

Leptin is a hormone made by AT physiologically acting primarily on neurons in the hypothalamus, regulating food intake and energy expenditure. The role of leptin in tumorigenesis was suggested by the high expression of its receptor (ObR) in several cancer cells. Both leptin and ObR were found overexpressed in BCs, especially in higher grade tumors, and were shown to be related with distant metastasis and poor prognosis [116]. Prevalently, leptin exerts its biological function through binding to its receptor, which activates multiple downstream signaling pathways such as STAT3, ERK, and PI3K signaling, involved in the control of cell proliferation, differentiation, survival, migration, and invasion [117]. Leptin signaling was also shown to be involved in the promotion of a stem-like phenotype related to resistance to therapy and tumor recurrence and metastasis [118]. Its pro-tumorigenic activity in BCs also derived from its interaction with the ER signaling. Indeed, leptin works on two levels; on one hand, it modulates estrogen production in adipose stroma, and on the other hand, it induces upregulation of ER expression and ER functional transactivation in BC cells governing estrogen sensitivity in cancer cells [119]. Accordingly, it has been shown that obese stromal cells producing large quantities of leptin increased proliferation and metastasis of several ER+BC cell lines through a leptin-mediated pathway [120]. Leptin was described to have a pro-tumorigenic role modulating the tumor immune microenvironment. Indeed, it was shown to induce STAT3 activation in tumor infiltrating effector T cells promoting FAO, which in turn facilitates the use of FAs as an energy source, impedes glycolysis associated with restricted CD8+ T cell antitumor function, and enhances BC progression [121]. Moreover, leptin has been described to mediate dysfunction of CD8+ T cells by increasing the expression of PD-1 [122]. These effects have clear implications in the efficacy of checkpoint inhibitors in obese patients.

Leptin is a key paracrine mediator regulating the interaction between stromal cells and BC cells in the meaning of tumor metabolism. It was shown that leptin released by AT contributes to the metabolic features associated with BC malignancy, such as switching the cells' energy balance from mitochondrial β-oxidation to the aerobic glycolytic pathway [123]. Obesity impacts leptin [77,78] and adiponectin levels in opposite manners, which means that not their absolute quantities but rather the mutual proportion in which they occur may be a key parameter indicating the relative risk of BC. Indeed, high leptin-to-adiponectin ratio is associated with increased risk of post-menopausal BC [124] and with increased progression in TNBC [125]. In addition, its expression characterized CAAs of BC tissue [20] and mammary fat tissue from BCs compared to fat derived from benign lesions [30].

5.3.2. Autotaxin and Resistin

Autotaxin (ATX) is a secreted glycoprotein produced by platelets, endothelial cells, fibroblast, adipocytes and, in varying degrees, by cancer cells [20]. Enzymatic activity of ATX converts lysophosphatidylcholine (LPC) into lipid signaling molecule lysophosphatidic acid (LPA), which controls key processes such as cell renewal, cell migration, proliferation, and survival [126]. Increased

ATX expression in tumors upregulates inflammatory programs and is associated with enhanced tumor progression, aggressiveness, increased angiogenesis, metastasis, and chemoresistance [127]. Benesch et al. [85] proposed a special role for the adipose compartment within the BC microenvironment in the supply of ATX. Namely, inflammatory signals secreted by tumor cells increase ATX expression in CAAs and fibroblasts, reinforcing the inflammatory vicious cycle and contributing to tumor progression. The pro-tumorigenic crosstalk between the tumor and the surrounding AT was shown to be interrupted by ATX inhibition. ATX is highly expressed in subcutaneous fat depots of obese patients [82] and is increased in circulation of patients with higher BMI [83,84].

Resistin is a fat-derived secretory factor usually found in inflammatory zones [128]. Recently published data evidenced that resistin promotes the metastatic potential of BC cells by inducing EMT and stemness [129]. As with other adipokines circulating in the blood, resistin may act on BC systemically and locally. It was found to be higher in the circulation of obese patients and ob/ob mice, a well-described model of obesity [67]. Also, CAAs in BC tissues produce this metastasis-inducing adipokine [20]. The resistin receptor adenylyl cyclase-associated protein 1 (CAP1) was shown to be expressed by numerous BC cell lines and primary human tumors. Moreover, its high expression in BC patients was associated with characteristics of aggressiveness and poor prognosis [130].

5.3.3. HGF and IGF

Adipocytes and preadipocytes are known to secret hepatocyte growth factor (HGF), which, together with its receptor present on tumor cells (c-Met), forms a signaling pathway intensely correlated with proliferation, metastasis, and angiogenesis [131]. HGF was shown to be highly secreted by adipocytes derived from obese individuals; accordingly, its serum level was higher in obese subjects when compared to lean subjects [86]. Of note, it was shown that c-Met expression was increased at the BC invasive front in adipocytes proximity. Although cancer cells seem to not increase expression of HGF in nearby adipocytes, the fact that expression of HGF and c-Met at the interface of adipocytes and cancer cells was correlated with histologic grade and reduced patient survival is certainly an important indication that the pathway initiated by the adipocytes-derived HGF has an impact on tumor progression [132].

Insulin-like growth factor 1 (IGF-1) is also secreted by adipocytes and preadipocytes, and its release is increased about two-fold in both undifferentiated cells and differentiated adipocytes from obese compared to lean individuals [87,88]. IGF-1, through the binding of its receptor IGF-1R expressed in tumor cells, activates PI3K/AKT and MAPK pathways, resulting in their enhanced proliferation. Suppression of IGF-1R in BC cells was shown to inhibit the tumor promoting effect of adipocytes.

5.4. Extracellular Matrix Remodeling

Cancer progression builds upon the ability of cancer cells to traverse the extracellular matrix (ECM) barrier access the circulation and establish distant metastases. The ECM composition not only orchestrates cancer and stromal cell behavior but also regulates the spectrum of reciprocal complex interactions between cells. It is widely recognized that both biochemical and biomechanical properties of the ECM influence cancer cell plasticity, allowing them to survive in hostile microenvironments and resist therapy. Production of ECM components functioning as signaling molecules and its continuous remodeling are clearly implicated in tumor progression. In AT, ECM is crucial for maintaining the structural integrity of adipocytes and plays a pivotal role in adipogenesis. Adipocytes are an abundant source of ECM components. However, tumor-induced adipocyte "activation" makes them produce even larger quantities of tumor promoting ECM components, such as collagen VI (COLVI) [19,21], which was described to contribute to AT fibrosis and inflammation in obesity [91,94]. COLVI was shown to promote tumor growth and survival signaling through the NG2/chondroitin sulfate proteoglycan receptor expressed on tumor cells [133]. The cleaved fragment of the COLVI α3 chain—endotrophin—acts a potent profibrotic factor stimulating TGF-β-dependent EMT [134].

In the same way, adipocytes at the invasive front of human tumors and CAAs in vitro exhibit increased expression of MMP11) [19,38]. MMP11 is a potent physiologic negative regulator of adipogenesis, and its expression in adipocytes was shown to be induced by tumor cells, in turn leading to the accumulation of non-malignant peritumoral fibroblast-like cells, promoting cancer cell survival and tumor progression. Thus, MMP11 plays the central role during tumor desmoplasia [135].

5.5. Metabolic Shift

Cancer cells invading regions of adipocytes in the tumor microenvironment interfere with essential functions of white AT, such as controlling energy balance, leading to the redistribution of nutrients in favor of cancer cells. It was shown that a tumor educates its stroma by downregulating p62, and such a prepared stroma shuts down energy utilization, which increases nutrient availability for cancer cells [136]. Recently, much attention has been paid to the role of adipocytes releasing FAs and other macronutrients supporting tumor growth [9,137]. Tumor cells switching from glycolysis to lipid-dependent energy production are supported by both de novo lipogenic synthesis and the acquisition of exogenous FAs [138]. FAs provide an excellent source of energy production through the FAO. Cancer cells can also store excess lipids in the form of lipid droplets, which supply energy to power their expansion and metastasis [138]. Especially, obese adipocytes supply more FAs to cancer cells than non-obese adipocytes, increasing the energy amount available for tumor growth and metastasis [34]. Cell surface fatty acid translocase (CD36) has been recognized as a marker of metastasis initiating cells in various cancer types, including BC [138,139]. Those cells are highly responsive to exogenous FAs, and the inhibition of CD36 was shown to impair metastasis [138,140]. FFAs released from CAAs serve not only as nutrients but can also be used for the biosynthesis of a series of lipid-signaling molecules that promote tumorigenesis [9].

Tumor-educated adipocytes due to strong metabolic pressure shift their metabolism to glycolysis with concomitant release of energy-rich metabolites, such as lactate and pyruvate. The dynamic monocarboxylate shuttle between BC cells and adipocytes through monocarboxylate transporters (MCT) plays a role in BC aggressiveness [141]. Indeed, it was shown that adipocytes co-cultivated with tumor cells exhibited upregulated expression of 4MCT4, facilitating lactic acid efflux [142]. Cancer cells, on the other hand, are equipped with an appropriate apparatus for lactic acid uptake, such as MCT1 and MCT2. In fact, highly proliferating ER-negative BC subtypes express high levels of MCT1, which is associated with poor patient outcome [141]. The association was even stronger for combined expressions of MCT4 and MCT1 in AT and tumor cells, respectively.

Ketone bodies produced and released by glycolytic adipocytes are an ideal substrate for ATP production by driving oxidative mitochondrial metabolism in invasive cancer. They may burn more efficiently than other mitochondrial fuels, even during hypoxia, potentially allowing the tumor to grow in the absence of an optimal blood supply [143]. Importantly, the coexistence of adipocytes and tumor cells potentiates both ketogenesis in adipocytes as well as ketolytic activity in BC cells [142]. Additionally, it was shown that β-hydroxybutyrate secreted from adipocytes enhanced BC cells malignancy in vitro, upregulating several tumor-promoting genes in BC cells [144]. Induction of ketone-specific gene signature was shown to be associated with worse outcomes in BC patients [145].

6. Adipocytes in Cancer-Associated Cachexia

Cancer-associated cachexia is a well-known complication of cancer that frequently is the cause of death in cancer patients. As a systemic illness, cachexia affects the vast majority of patients with end stage cancer. Due to an energy imbalance condition and inflammation, the body burns its own components to sustain the tumor growth. Cachexia differs from malnutrition in as much as cachexia cannot be successfully treated by supplemental nutrition alone [145]. Cachexia has been most frequently recognized in the course of pancreatic, gastric, colorectal, lung, head, and neck cancers [146]. Although in the past cachexia was not often documented in BC patients, more recent studies suggest its particular impact on bone metastasis common in patients with advanced BC [147]. The previously rather neglected AT dysfunction is an essential contributor to cachexia and has been recognized as an early phenomenon occurring just before the skeletal muscle atrophy [148,149]. This is in line with observations carried out in tumor-bearing mice, which confirmed that AT wasting is an early event occurring at a time when the tumor is hardly palpable [150]. Moreover, the breakdown of adipocyte triacylglycerols (which is at the basis of cachexia-associated fat tissue remodeling) may actually activate muscle proteolysis [151]. Several serum factors such as TNF-α, IL-1β, IL-6, and a zinc-glycoprotein (ZAG), also called lipid-mobilizing factors secreted by tumor or host cells, have been shown to be involved in local as well as systemic AT lipolysis [145]. Increased FAs levels in circulation can be taken up by skeletal muscle, and the excess of intramuscular FAs was shown to cause several biochemical changes, such as the expression of lipases Atrogin-1 and MuRF67, leading to skeletal muscle atrophy [152,153]. On the other hand, skeletal muscle atrophy may act as positive feedback enhancing AT lipolysis [154]. Thus, the particular crosstalk between AT and muscles in cachexia is an undeniable fact and contributes to the progression of cachexia.

Cachectic adipocytes under some aspects, such as morphologic modifications, decrease in lipid content [25], decrease in expression of C/EBPα [27] and leptin [79], as well as proinflammatory phenotype [65,66,72] and induction of ECM remodeling [27,65,93], resemble CAAs [155] (Tables 1 and 3). Also, characteristics of cancer cachexia (i.e., decreased glucose metabolism along with decreased lipogenic rate [32]) were observed in CAAs. There are also important features of cachectic adipocytes not seen in CAAs, such as upregulation of lipolytic enzymes (*HSL* or *ATGL*) [35], making them similar to adipocytes in obesity. During AT lipolysis, ATGL and HSL hydrolyze stored triglycerides and produce FFAs and glycerol. In line with this, cachectic—similarly to obese patients—manifests high levels of circulating FFAs, glycerol, and triacylglycerol [156–158].

A very reasonable question arises, namely, whether excess body fat and obesity can contribute to cancer cachexia occurrence and worsening patient prognosis. As (i) illnesses associated with insulin resistance, such as obesity, exhibit increases in whole-body protein degradation, (ii) insulin resistance is associated with muscle fat accumulation, and (iii) high-fat diet enhances muscle protein catabolism associated with an increase in plasma FAs and a decrease in plasma adiponectin, the implication of obesity in cancer cachexia is an entirely plausible scenario [159]. Body composition, defined as the proportions and the distribution of lean and fat tissues, is an emergent issue in clinical oncology. Occurrence of sarcopenia, a severe muscle depletion, is most easily missed in obese patients. Coexistence of obesity and sarcopenia, termed "sarcopenic obesity", are highly prevalent [160]. Mechanisms of muscle loss observed in patients with sarcopenic obesity may involve AT homeostasis and thus likely overlap with mechanisms of cancer cachexia. Obesity was shown to be associated with a decrease in circulating adiponectin and increased IL-6 and FFAs levels (Table 3). These futures were shown to trigger muscle protein loss with concomitant accumulation of fat in muscle fibers [159].

Tumor cells are highly energy-demanding, thus energy and metabolic intermediates such as FFAs released from AT in both obese and cachectic are desired substrates required to sustain their proliferation. However, certain cancers, along with high rate lipolysis, were shown to induce white-to-brown transdifferentiation, named "browning", mediated by UCP1, resulting in enhanced fat energy burning with concomitant fat tissue wasting [31]. Paradoxically, this means that energy that could be consumed by tumor cells is burned by thermogenic reactions in adipocytes. In the murine prostate cancer model, p62 emerged as the main regulator of adipocyte metabolic health (supporting browning and adipogenesis). p62 was shown to guard the fat–tumor interaction, as its loss in adipocytes severely inhibited adipocyte browning, allowing establishment of a symbiotic interaction based on adipocyte low energy utilization in favor of tumor needs. In fact, also in BC, p62 levels were reduced in the stroma of several tumors, and its loss resulted in increased tumorigenesis [161].

The process of cancer induced adipocyte browning is largely described in cancer cachexia and concerns adipocytes in the whole body, but this phenomenon has been also observed in the tumor microenvironment. Indeed, higher browning of adipocytes was noted in AT adjacent to BCs when compared to benign lesions [30]. Highly UCP1 expressing cancer-associated fibroblasts were shown to have tumor promoting effects via the generation of high-energy mitochondrial fuels (such as ketone bodies). In fact, this means that even activation of the stroma towards energy consumption (browning) may have a pro-tumorigenic effect through the production of alternative and efficient energy substrates. In general, both adipocytes activation in the tumor vicinity as well as adipocytes extreme modifications observed in cancer cachexia, together with their similarities and differences, are certainly not negligible complications of tumor progression, as cancer cells induce and exploit local and systemic functions for its purpose.

While it seems possible that restoring AT homeostasis by targeting these areas of dysfunction would contribute to further improvements in patient-centered outcomes, current clinical trials investigating potential therapeutic agents for patients with cancer associated cachexia do not directly address the contributions of white, beige, and brown AT to cancer cachexia [154].

7. Conclusions

Adipocytes are active players in the development and the progression of BCs. Through the secretion of several stimuli, they create a permissive microenvironment for tumor growth. In the current review, we compared adipocytes involved in three pathologies—obesity, BC, and cachexia. In our opinion, such a comparison may help to understand the way adipocytes support tumor growth and, above all, it can hopefully bring us closer to understanding how cancer cells introduce modifications in AT not found in normal AT. Adipocyte characteristics in obesity are similar to those detected in CAAs of invasive tumors, raising the possibility that adipocytes support tumor development and progression in obese people due to their already altered phenotype. At first glance, such different adipocytes observed in obesity and cachexia turned out to be very similar, especially in terms of metabolic disturbances and pro-inflammatory features, hence becoming dangerous partners of tumor progression. Importantly, altered phenotypes of CAAs and cachectic adipocytes are the results of tumor direct/indirect actions, local and systemic, respectively. Indeed, they both have detrimental effects when it comes to tumor progression and patient prognosis.

Adipocytes, as active player in tumor progression, are emerging as a new target in treating BC. New drugs can be developed to target adipocytes and/or cancer cells to block the adipocyte–tumor cells' vicious cycle from inducing cancer progression. In this context, further research is required to understand the mechanism(s) by which tumor cells modify adipocytes. Tumor-derived signals to adipocytes might be those used by tumor cells to continuously communicate with and activate stromal cells in the bone marrow or in distant organs in order to progress into metastasis. Moreover, blocking such signals could help in countering tumor-derived adipocyte dysfunction such as cachexia. On the other hand, understanding the efficacy of adipocyte-derived factor inhibition on cancer progression could represent a new strategy to treat cancer, especially in obese patients. Curiously, adipocytes have been also exploited to be the final result of a trans-differentiation process induced in BC cells to reduce BC progression [162]. Indeed, it was recently demonstrated that cancer cells undergoing an EMT program to adapt to changing signals from the microenvironment and to escape drug treatments acquired cellular plasticity that could be exploited to force trans-differentiation of BC cells into bona fide postmitotic adipocytes. This treatment was demonstrated to inhibit cancer metastasis in several preclinical models. Based on the relevance of the interaction between tumor cells and adipocytes in BC progression reviewed above, a potential contribution of the cancer cell-derived adipocytes to tumor progression warrants further investigation before the use of this strategy as a therapeutic approach.

Funding: This work was supported by Associazione Italiana Ricerca sul Cancro (AIRC), (No. 18712) (TT).

Acknowledgments: We thank Mameli L. for secretarial assistance.

Conflicts of Interest: The authors declare no conflicts of interest.

References

1. Tan, J.; Buache, E.; Chenard, M.P.; Dali-Youcef, N.; Rio, M.C. Adipocyte is a non-trivial, dynamic partner of breast cancer cells. *Int. J. Dev. Biol.* **2011**, *55*, 851–859. [CrossRef] [PubMed]
2. Hotamisligil, G.S. Inflammation, metaflammation and immunometabolic disorders. *Nature* **2017**, *542*, 177–185. [CrossRef] [PubMed]
3. Zhang, Y.; Proenca, R.; Maffei, M.; Barone, M.; Leopold, L.; Friedman, J. Positional cloning of the mouse obese gene and its human homologue. *Nature* **1994**, *372*, 425–432. [CrossRef] [PubMed]
4. Coelho, M.; Oliveira, T.; Fernandes, R. Biochemistry of adipose tissue: An endocrine organ. *Arch. Med. Sci.* **2013**, *9*, 191–200. [CrossRef] [PubMed]
5. Ouchi, N.; Parker, J.L.; Lugus, J.J.; Walsh, K. Adipokines in inflammation and metabolic disease. *Nat. Rev. Immunol.* **2011**, *11*, 85–97. [CrossRef] [PubMed]
6. Proebstle, T.M.; Huber, R.; Sterry, W. Detection of early micrometastases in subcutaneous fat of primary malignant melanoma patients by identification of tyrosinase-mRNA. *Eur. J. Cancer* **1996**, *32A*, 1664–1667. [CrossRef]
7. Hardaway, A.L.; Herroon, M.K.; Rajagurubandara, E.; Podgorski, I. Bone marrow fat: Linking adipocyte-induced inflammation with skeletal metastases. *Cancer Metastasis Rev.* **2014**, *33*, 527–543. [CrossRef]
8. Wu, Q.; Sun, S.; Li, Z.; Yang, Q.; Li, B.; Zhu, S.; Wang, L.; Wu, J.; Yuan, J.; Wang, C.; et al. Breast cancer-released exosomes trigger cancer-associated cachexia to promote tumor progression. *Adipocyte* **2019**, *8*, 31–45.
9. Martinez-Outschoorn, U.; Sotgia, F.; Lisanti, M.P. Tumor microenvironment and metabolic synergy in breast cancers: Critical importance of mitochondrial fuels and function. *Semin. Oncol.* **2014**, *41*, 195–216. [CrossRef]
10. Ghaben, A.L.; Scherer, P.E. Adipogenesis and metabolic health. *Nat. Rev. Mol. Cell Biol.* **2019**, *20*, 242–258. [CrossRef]
11. Rosen, E.D.; Walkey, C.J.; Puigserver, P.; Spiegelman, B.M. Transcriptional regulation of adipogenesis. *Genes Dev.* **2000**, *14*, 1293–1307.
12. Ibrahim, M.M. Subcutaneous and visceral adipose tissue: Structural and functional differences. *Obes. Rev.* **2010**, *11*, 11–18. [CrossRef] [PubMed]
13. Neeland, I.J.; Ayers, C.R.; Rohatgi, A.K.; Turer, A.T.; Berry, J.D.; Das, S.R.; Vega, G.L.; Khera, A.; McGuire, D.K.; Grundy, S.M.; et al. Associations of visceral and abdominal subcutaneous adipose tissue with markers of cardiac and metabolic risk in obese adults. *Obesity* **2013**, *21*, E439–E447. [CrossRef] [PubMed]
14. Hovey, R.C.; Aimo, L. Diverse and active roles for adipocytes during mammary gland growth and function. *J. Mammary Gland Biol. Neoplasia* **2010**, *15*, 279–290. [CrossRef] [PubMed]
15. Wang, Q.A.; Song, A.; Chen, W.; Schwalie, P.C.; Zhang, F.; Vishvanath, L.; Jiang, L.; Ye, R.; Shao, M.; Tao, C.; et al. Reversible De-differentiation of Mature White Adipocytes into Preadipocyte-like Precursors during Lactation. *Cell Metab.* **2018**, *28*, 282–288. [CrossRef] [PubMed]
16. Morroni, M.; Giordano, A.; Zingaretti, M.C.; Boiani, R.; De Matteis, R.; Kahn, B.B.; Nisoli, E.; Tonello, C.; Pisoschi, C.; Luchetti, M.M.; et al. Reversible transdifferentiation of secretory epithelial cells into adipocytes in the mammary gland. *Proc. Natl. Acad. Sci. USA* **2004**, *101*, 16801–16806. [CrossRef] [PubMed]
17. Cinti, S. Pink Adipocytes. *Trends Endocrinol. Metab.* **2018**, *29*, 651–666. [CrossRef]
18. Zwick, R.K.; Rudolph, M.C.; Shook, B.A.; Holtrup, B.; Roth, E.; Lei, V.; Van Keymeulen, A.; Seewaldt, V.; Kwei, S.; Wysolmerski, J.; et al. Adipocyte hypertrophy and lipid dynamics underlie mammary gland remodeling after lactation. *Nat. Commun.* **2018**, *9*, 3592. [CrossRef]
19. Dirat, B.; Bochet, L.; Dabek, M.; Daviaud, D.; Dauvillier, S.; Majed, B.; Wang, Y.Y.; Meulle, A.; Salles, B.; Le, G.S.; et al. Cancer-associated adipocytes exhibit an activated phenotype and contribute to breast cancer invasion. *Cancer Res.* **2011**, *71*, 2455–2465. [CrossRef]
20. Choi, J.; Cha, Y.J.; Koo, J.S. Adipocyte biology in breast cancer: From silent bystander to active facilitator. *Prog. Lipid Res.* **2018**, *69*, 11–20. [CrossRef]
21. Nieman, K.M.; Kenny, H.A.; Penicka, C.V.; Ladanyi, A.; Buell-Gutbrod, R.; Zillhardt, M.R.; Romero, I.L.; Carey, M.S.; Mills, G.B.; Hotamisligil, G.S.; et al. Adipocytes promote ovarian cancer metastasis and provide energy for rapid tumor growth. *Nat. Med.* **2011**, *17*, 1498–1503. [CrossRef] [PubMed]

22. Bochet, L.; Lehud, C.; Dauvillier, S.; Wang, Y.Y.; Dirat, B.; Laurent, V.; Dray, C.; Guiet, R.; Maridonneau-Parini, I.; Le Gonidec, S.; et al. Adipocyte-derived fibroblasts promote tumor progression and contribute to the desmoplastic reaction in breast cancer. *Cancer Res.* **2013**, *73*, 5657–5668. [CrossRef] [PubMed]

23. Ma, S.; Jing, F.; Xu, C.; Zhou, L.; Song, Y.; Yu, C.; Jiang, D.; Gao, L.; Li, Y.; Guan, Q.; et al. Thyrotropin and obesity: Increased adipose triglyceride content through glycerol-3-phosphate acyltransferase 3. *Sci. Rep.* **2015**, *5*, 7633. [CrossRef] [PubMed]

24. Cai, Z.; Liang, Y.; Xing, C.; Wang, H.; Hu, P.; Li, J.; Huang, H.; Wang, W.; Jiang, C. Cancer-associated adipocytes exhibit distinct phenotypes and facilitate tumor progression in pancreatic cancer. *Oncol. Rep.* **2019**, *42*, 2537–2549. [CrossRef] [PubMed]

25. Dalal, S. Lipid metabolism in cancer cachexia. *Ann. Palliat. Med.* **2019**, *8*, 13–23. [CrossRef]

26. Boughanem, H.; Cabrera-Mulero, A.; Millán-Gómez, M.; Garrido-Sánchez, L.; Cardona, F.; Tinahones, F.J.; Moreno-Santos, I.; Macías-González, M. Transcriptional Analysis of FOXO1, C/EBP-α and PPAR-γ2 Genes and Their Association with Obesity-Related Insulin Resistance. *Genes* **2019**, *10*, 706. [CrossRef]

27. Bing, C.; Russell, S.; Becket, E.; Pope, M.; Tisdale, M.J.; Trayhurn, P.; Jenkins, J.R. Adipose atrophy in cancer cachexia: Morphologic and molecular analysis of adipose tissue in tumour-bearing mice. *Br. J. Cancer* **2006**, *95*, 1028–1037. [CrossRef]

28. Winn, N.C.; Vieira-Potter, V.J.; Gastecki, M.L.; Welly, R.J.; Scroggins, R.J.; Zidon, T.M.; Gaines, T.L.; Woodford, M.L.; Karasseva, N.G.; Kanaley, J.A.; et al. Loss ofUCP1 exacerbates Western diet-induced glycemic dysregulation independent of changes in body weight in female mice. *Am. J. Physiol. Regul. Integr. Comp. Physiol.* **2017**, *312*, R74–R84. [CrossRef]

29. Fromme, T.; Klingenspor, M. Uncoupling protein 1 expression and high-fat diets. *Am. J. Physiol. Regul. Integr. Comp. Physiol.* **2011**, *300*, R1–R8. [CrossRef]

30. Wang, F.; Gao, S.; Chen, F.; Fu, Z.; Yin, H.; Lu, X.; Yu, J.; Lu, C. Mammary fat of breast cancer: Gene expression profiling and functional characterization. *PLoS ONE* **2014**, *9*, e109742. [CrossRef]

31. Kir, S.; Spiegelman, B.M. Cachexia & brown fat: A burning issue in cancer. *Trends Cancer* **2016**, *2*, 461–463. [PubMed]

32. Kir, S.; White, J.P.; Kleiner, S.; Kazak, L.; Cohen, P.; Baracos, V.E.; Spiegelman, B.M. Tumour-derived PTH-related protein triggers adipose tissue browning and cancer cachexia. *Nature* **2014**, *513*, 100–104. [CrossRef] [PubMed]

33. Lewandowski, P.A.; Cameron-Smith, D.; Jackson, C.J.; Kultys, E.R.; Collier, G.R. The role of lipogenesis in the development of obesity and diabetes in Israeli sand rats (Psammomys obesus). *J. Nutr.* **1998**, *128*, 1984–1988. [CrossRef]

34. Balaban, S.; Shearer, R.F.; Lee, L.S.; van Geldermalsen, M.; Schreuder, M.; Shtein, H.C.; Cairns, R.; Thomas, K.C.; Fazakerley, D.J.; Grewal, T.; et al. Adipocyte lipolysis links obesity to breast cancer growth: Adipocyte-derived fatty acids drive breast cancer cell proliferation and migration. *Cancer Metab.* **2017**, *13*, 5. [CrossRef]

35. Agustsson, T.; Rydén, M.; Hoffstedt, J.; van Harmelen, V.; Dicker, A.; Laurencikiene, J.; Isaksson, B.; Permert, J.; Arner, P. Mechanism of increased lipolysis in cancer cachexia. *Cancer Res.* **2007**, *67*, 5531–5537. [CrossRef]

36. Tejerina, S.; De Pauw, A.; Vankoningsloo, S.; Houbion, A.; Renard, P.; De Longueville, F.; Raes, M.; Arnould, T. Mild mitochondrial uncoupling induces 3T3-L1 adipocyte de-differentiation by a PPARgamma-independent mechanism, whereas TNFalpha-induced de-differentiation is PPARgamma dependent. *J. Cell Sci.* **2009**, *122*, 145–155. [CrossRef]

37. Zoico, E.; Darra, E.; Rizzatti, V.; Budui, S.; Franceschetti, G.; Mazzali, G.; Rossi, A.P.; Fantin, F.; Menegazzi, M.; Cinti, S.; et al. Adipocytes WNT5a mediated dedifferentiation: A possible target in pancreatic cancer microenvironment. *Oncotarget* **2016**, *7*, 20223–20235. [CrossRef]

38. Andarawewa, K.L.; Motrescu, E.R.; Chenard, M.P.; Gansmuller, A.; Stoll, I.; Tomasetto, C.; Rio, M.C. Stromelysin-3 is a potent negative regulator of adipogenesis participating to cancer cell-adipocyte interaction/crosstalk at the tumor invasive front. *Cancer Res.* **2005**, *65*, 10862–10871. [CrossRef] [PubMed]

39. Burton, B.T.; Foster, W.R.; Hirsch, J.; Van Itallie, T.B. Health implications of obesity: An NIH Consensus Development Conference. *Int J. Obes* **1985**, *9*, 155–170. [PubMed]

40. Ng, M.; Fleming, T.; Robinson, M.; Thomson, B.; Graetz, N.; Margono, C.; Mullany, E.C.; Biryukov, S.; Abbafati, C.; Abera, S.F.; et al. Global, regional, and national prevalence of overweight and obesity in children and adults during 1980–2013: A systematic analysis for the Global Burden of Disease Study 2013. *Lancet* **2014**, *384*, 766–781. [CrossRef]

41. NCD Risk Factor Collaboration (NCD-RisC) Trends in adult body-mass index in 200 countries from 1975 to 2014: A pooled analysis of 1698 population-based measurement studies with 19-+2 million participants. *Lancet* **2016**, *387*, 1377–1396. [CrossRef]

42. Siegel, R.L.; Miller, K.D.; Jemal, A. Cancer statistics, 2019. *CA Cancer J. Clin.* **2019**, *69*, 7–34. [CrossRef] [PubMed]

43. Kerr, J.; Anderson, C.; Lippman, S.M. Physical activity, sedentary behaviour, diet, and cancer: An update and emerging new evidence. *Lancet Oncol.* **2017**, *18*, e457–e471. [CrossRef]

44. Masala, G.; Bendinelli, B.; Assedi, M.; Occhini, D.; Zanna, I.; Sieri, S.; Agnoli, C.; Sacerdote, C.; Riccardi, F.; Mattiello, A.; et al. Up to one-third of breast cancer cases in post-menopausal Mediterranean women might be avoided by modifying lifestyle habits: The EPIC Italy study. *Breast Cancer Res. Treat.* **2017**, *161*, 311–320. [CrossRef]

45. Renehan, A.G.; Tyson, M.; Egger, M.; Heller, R.F.; Zwahlen, M. Body-mass index and incidence of cancer: A systematic review and meta-analysis of prospective observational studies. *Lancet* **2008**, *371*, 569–578. [CrossRef]

46. Eliassen, A.H.; Coldlitz, G.A.; Rosner, B.; Willet, W.C.; Hankinson, S.E. Adult weight change and risk of postmenopausal breast cancer. *JAMA* **2006**, *296*, 193–201.

47. McKenzie, F.; Ferrari, P.; Freisling, H.; Chajes, V.; Rinaldi, S.; de Batlle, J.; Dahm, C.C.; Overvad, K.; Baglietto, L.; Dartois, L.; et al. Healthy lifestyle and risk of breast cancer among postmenopausal women in the European Prospective Investigation into Cancer and Nutrition cohort study. *Int. J. Cancer* **2015**, *136*, 2640–2648.

48. Vrieling, A.; Buck, K.; Kaaks, R.; Chang-Claude, J. Adult weight gain in relation to breast cancer risk by estrogen and progesterone receptor status: A meta-analysis. *Breast Cancer Res. Treat.* **2010**, *123*, 641–649.

49. Yang, X.R.; Chang-Claude, J.; Goode, E.L.; Couch, F.J.; Nevanlinna, H.; Milne, R.L.; Gaudet, M.; Schmidt, M.K.; Broeks, A.; Cox, A.; et al. Associations of breast cancer risk factors with tumor subtypes: A pooled analysis from the Breast Cancer Association Consortium studies. *J. Natl. Cancer Inst.* **2011**, *103*, 250–263.

50. Suzuki, R.; Orsini, N.; Saji, S.; Key, T.J.; Wolk, A. Body weight and incidence of breast cancer defined by estrogen and progesterone receptor status—A meta-analysis. *Int. J. Cancer* **2009**, *124*, 698–712. [CrossRef]

51. Pierobon, M.; Frankenfeld, C.L. Obesity as a risk factor for triple-negative breast cancers: A systematic review and meta-analysis. *Breast Cancer Res. Treat.* **2013**, *137*, 307–314. [CrossRef] [PubMed]

52. Protani, M.; Coory, M.; Martin, J.H. Effect of obesity on survival of women with breast cancer: Systematic review and meta-analysis. *Breast Cancer Res. Treat.* **2010**, *123*, 627–635. [CrossRef] [PubMed]

53. Chan, D.S.M.; Vieira, A.R.; Aune, D.; Bandera, E.V.; Greenwood, D.C.; McTiernan, A.; Navarro Rosenblatt, D.; Thune, I.; Vieira, R.; Norat, T. Body mass index and survival in women with breast cancer-systematic literature review and meta-analysis of 82 follow-up studies. *Ann. Oncol.* **2014**, *25*, 1901–1914. [CrossRef] [PubMed]

54. Playdon, M.C.; Bracken, M.B.; Sanft, T.B.; Ligibel, J.A.; Harrigan, M.; Irwin, M.L. Weight gain after breast cancer diagnosis and all-cause mortality: Systematic review and meta-analysis. *J. Natl. Cancer Inst.* **2015**, *107*, djv275. [CrossRef] [PubMed]

55. Majed, B.; Dozol, A.; Ribassin-Majed, L.; Senouci, K.; Asselain, B. Increased risk of contralateral breast cancers among overweight and obese women: A time-dependent association. *Breast Cancer Res. Treat.* **2011**, *126*, 729–738. [CrossRef] [PubMed]

56. Ewertz, M.; Jensen, M.B.; Gunnarsdottir, K.; Hoiris, I.; Jakobsen, E.H.; Nielsen, D.; Stenbygaard, L.E.; Tange, U.B.; Cold, S. Effect of obesity on prognosis after early-stage breast cancer. *J. Clin. Oncol.* **2011**, *29*, 25–31. [CrossRef]

57. Majed, B.; Moreau, T.; Senouci, K.; Salmon, R.J.; Fourquet, A.; Asselain, B. Is obesity an independent prognosis factor in woman breast cancer? *Breast Cancer Res. Treat.* **2008**, *111*, 329–342. [CrossRef]

58. Yamaguchi, J.; Ohtani, H.; Nakamura, K.; Shimokawa, I.; Kanematsu, T. Prognostic impact of marginal adipose tissue invasion in ductal carcinoma of the breast. *Am. J. Clin. Pathol.* **2008**, *130*, 382–388. [CrossRef]

59. Kimijima, I.; Ohtake, T.; Sagara, H.; Watanabe, T.; Takenoshita, S. Scattered fat invasion: An indicator for poor prognosis in premenopausal, and for positive estrogen receptor in postmenopausal breast cancer patients. *Oncology* **2000**, *59*, 25–30. [CrossRef]

60. Lee, Y.; Jung, W.H.; Koo, J.S. Adipocytes can induce epithelial-mesenchymaltransition in breast cancer cells. *Breast Cancer Res. Treat.* **2015**, *153*, 323–335. [CrossRef]

61. Manabe, Y.; Toda, S.; Miyazaki, K.; Sugihara, H. Mature adipocytes, but not preadipocytes, promote the growth of breast carcinoma cells in collagen gel matrix culture through cancer-stromal cell interactions. *J. Pathol.* **2003**, *201*, 221–228. [CrossRef]

62. Lim, S.; Hosaka, K.; Nakamura, M.; Cao, Y. Co-option of pre-existing vascular beds in adipose tissue controls tumor growth rates and angiogenesis. *Oncotarget* **2016**, *7*, 38282–38291. [CrossRef] [PubMed]

63. Schweiger, M.; Eichmann, T.O.; Taschler, U.; Zimmermann, R.; Zechner, R.; Lass, A. Measurement of lipolysis. *Methods Enzymol.* **2014**, *538*, 171–193. [PubMed]

64. Zahorska-Markiewicz, B.; Janowska, J.; Olszanecka-Glinianowicz, M.; Zurakowski, A. Serum concentrations of TNF-alpha and soluble TNF-alpha receptors in obesity. *Int. Obes. Relat. Metab. Disord.* **2000**, *24*, 1392–1395. [CrossRef]

65. Alves, M.J.; Figuerêdo, R.G.; Azevedo, F.F.; Cavallaro, D.A.; Neto, N.I.; Lima, J.D.; Matos-Neto, E.; Radloff, K.; Riccardi, D.M.; Camargo, R.G.; et al. Adipose tissue fibrosis in human cancer cachexia: The role of TGFβ pathway. *BMC Cancer* **2017**, *17*, 190. [CrossRef]

66. de Matos-Neto, E.M.; Lima, J.D.; de Pereira, W.O.; Figuerêdo, R.G.; Riccardi, D.M.; Radloff, K.; das Neves, R.X.; Camargo, R.G.; Maximiano, L.F.; Tokeshi, F.; et al. Systemic Inflammation in Cachexia – Is Tumor Cytokine Expression Profile the Culprit? *Front. Immunol.* **2015**, *6*, 629. [CrossRef]

67. Akki, K.; Froguel, P.; Wolowczuk, I. Adipose tissue in obesity-related inflammation and insulin resistance: Cells, cytokines, and chemokines. *ISRN Inflamm.* **2013**, *2013*, 139239.

68. Bastard, J.P.; Jardel, C.; Bruckert, E.; Blondy, P.; Capeau, J.; Laville, M.; Vidal, H.; Hainque, B. Elevated levels of interleukin 6 are reduced in serum and subcutaneous adipose tissue of obese women after weight loss. *J. Clin. Endocrinol. Metab.* **2000**, *85*, 3338–3342.

69. Moschen, A.R.; Molnar, C.; Enrich, B.; Geiger, S.; Ebenbichler, C.F.; Tilg, H. Adipose and liver expression of interleukin (IL)-1 family members in morbid obesity and effects of weight loss. *Mol. Med.* **2011**, *17*, 840–845. [CrossRef]

70. Chen, A.; Mumick, S.; Zhang, C.; Lamb, J.; Dai, H.; Weingarth, D.; Mudgett, J.; Chen, H.; MacNeil, D.J.; Reitman, M.L.; et al. Diet induction of monocyte chemoattractant protein-1 and its impact on obesity. *Obes. Res.* **2005**, *13*, 1311–1320. [CrossRef]

71. Fujisaki, K.; Fujimoto, H.; Sangai, T.; Nagashima, T.; Sakakibara, M.; Shiina, N.; Kuroda, M.; Aoyagi, Y.; Miyazaki, M. Cancer-mediated adipose reversion promotes cancer cell migration via IL-6 and MCP-1. *Breast Cancer Res. Treat.* **2015**, *150*, 255–263. [CrossRef] [PubMed]

72. Talbert, E.E.; Lewis, H.L.; Farren, M.R.; Ramsey, M.L.; Chakedis, J.M.; Rajasekera, P.; Haverick, E.; Sarna, A.; Bloomston, M.; Pawlik, T.M.; et al. Circulating monocyte chemoattractant protein-1 (MCP-1) is associated with cachexia in treatment-naïve pancreatic cancer patients. *J. Cachexia Sarcopenia Muscle* **2018**, *9*, 358–368. [CrossRef] [PubMed]

73. Keophiphath, M.; Rouault, C.; Divoux, A.; Clément, K.; Lacasa, D. CCL5 promotes macrophage recruitment and survival in human adipose tissue. *Arterioscler. Thromb. Vasc. Biol.* **2010**, *30*, 39–45. [CrossRef]

74. D'Esposito, V.; Liguoro, D.; Ambrosio, M.R.; Collina, F.; Cantile, M.; Spinelli, R.; Raciti, G.A.; Miele, C.; Valentino, R.; Campiglia, P.; et al. Adipose microenvironment promotes triple negative breast cancer cell invasiveness and dissemination by producing CCL5. *Oncotarget* **2016**, *7*, 24495–24509. [CrossRef]

75. Morris, P.G.; Hudis, C.A.; Giri, D.; Morrow, M.; Falcone, D.J.; Zhou, X.K.; Du, B.; Brogi, E.; Crawford, C.B.; Kopelovich, L.; et al. Inflammation and increased aromatase expression occur in the breast tissue of obese women with breast cancer. *Cancer Prev. Res. (Phila)* **2011**, *4*, 1021–1029. [CrossRef]

76. O'Neill, J.S.; Elton, R.A.; Miller, W.R. Aromatase activity in adipose tissue from breast quadrants: A link with tumour site. *Br. Med. J. (Clin. Res. Ed.)* **1988**, *296*, 741–743. [CrossRef]

77. Hamilton, B.S.; Paglia, D.; Kwan, A.Y.; Deitel, M. Increased obese mRNA expression in omental fat cells from massively obese humans. *Nat. Med.* **1995**, *1*, 953–956. [CrossRef]

78. Considine, R.V.; Sinha, M.K.; Heiman, M.L.; Kriauciunas, A.; Stephens, T.W.; Nyce, M.R.; Ohannesian, J.P.; Marco, C.C.; McKee, L.J.; Bauer, T.L. Serum immunoreactive-leptin concentrations in normal-weight and obese humans. *N. Engl. J. Med.* **1996**, *334*, 292–295. [CrossRef]

79. Engineer, D.R.; Garcia, J.M. Leptin in anorexia and cachexia syndrome. *Int. J. Pept.* **2012**, *2012*, 287457. [CrossRef]

80. Cnop, M.; Havel, P.J.; Utzschneider, K.M.; Carr, D.B.; Sinha, M.K.; Boyko, E.J.; Retzlaff, B.M.; Knopp, R.H.; Brunzell, J.D.; Kahn, S.E. Relationship of adiponectin to body fat distribution, insulin sensitivity and plasma lipoproteins: Evidence for independent roles of age and sex. *Diabetologia* **2003**, *46*, 459–469. [CrossRef]

81. Wolf, I.; Sadetzki, S.; Kanety, H.; Kundel, Y.; Pariente, C.; Epstein, N.; Oberman, B.; Catane, R.; Kaufman, B.; Shimon, I. Adiponectin, ghrelin, and leptin in cancer cachexia in breast and colon cancer patients. *Cancer* **2006**, *106*, 966–973. [CrossRef] [PubMed]

82. Rancoule, C.; Dusaulcy, R.; Tréguer, K.; Grès, S.; Guigné, C.; Quilliot, D.; Valet, P.; Saulnier-Blache, J.S. Depot-specific regulation of autotaxin with obesity in human adipose tissue. *J. Physiol. Biochem.* **2012**, *68*, 635–644. [CrossRef] [PubMed]

83. Ferry, G.; Tellier, E.; Try, A.; Gres, S.; Naime, I.; Simon, M.F.; Rodriguez, M.; Boucher, J.; Tack, I.; Gesta, S.; et al. Autotaxin is released from adipocytes, catalyzes lysophosphatidic acid synthesis, and activates preadipocyte proliferation. Up-regulated expression with adipocyte differentiation and obesity. *J. Biol. Chem.* **2003**, *278*, 18162–18169. [CrossRef] [PubMed]

84. Boucher, J.; Quilliot, D.; Praderes, J.P.; Simon, M.F.; Gres, S.; Guigne, C.; Prévot, D.; Ferry, G.; Boutin, J.A.; Carpéné, C.; et al. Potential involvement of adipocyte insulin resistance in obesity-associated up-regulation of adipocyte lysophospholipase D/autotaxin expression. *Diabetologia* **2005**, *48*, 569–577. [CrossRef] [PubMed]

85. Benesch, M.G.K.; Tang, X.; Dewald, J.; Dong, W.F.; Mackey, J.R.; Hemmings, D.G.; McMullen, T.P.W.; Brindley, D.N. Tumor-induced inflammation in mammary adipose tissue stimulates a vicious cycle of autotaxin expression and breast cancer progression. *FASEB J.* **2015**, *29*, 3990–4000. [CrossRef] [PubMed]

86. Bell, L.N.; Ward, J.L.; Degawa-Yamauchi, M.; Bovenkerk, J.E.; Jones, R.; Cacucci, B.M.; Gupta, C.E.; Sheridan, C.; Sheridan, K.; Shankar, S.S.; et al. Adipose tissue production of hepatocyte growth factor contributes to elevated serum HGF in obesity. *Am. J. Physiol. Endocrinol. Metab.* **2006**, *291*, E843–E848. [CrossRef]

87. Nam, S.Y.; Lee, E.J.; Kim, K.R.; Cha, B.S.; Song, Y.D.; Lim, S.K.; Lee, H.C.; Huh, K.B. Relationship to IGF-binding protein (BP)-1, IGFBP-2, IGFBP-3, insulin, and growth hormone. *Int J. Obes. Relat. Metab. Disord.* **1997**, *21*, 355–359. [CrossRef]

88. D'Esposito, V.; Passaretti, F.; Hammarstedt, A.; Liguoro, D.; Terracciano, D.; Molea, G.; Canta, L.; Miele, C.; Smith, U.; Beguinot, F.; et al. Adipocyte-released insulin-like growth factor-1 is regulated by glucose and fatty acids and controls breast cancer cell growth in vitro. *Diabetologia* **2012**, *55*, 2811–2822. [CrossRef]

89. Hockenbery, D.; Nunez, G.; Milliman, C.; Schreiber, R.D.; Korsmeyer, S.J. Bcl-2 is an inner mitochondrial membrane protein that blocks programmed cell death. *Nature* **1990**, *348*, 334–336. [CrossRef]

90. Landin, K.; Stigendal, L.; Eriksson, E.; Krotkiewski, M.; Risberg, B.; Tengborn, L.; Smith, U. Abdominal obesity is associated with an impaired fibrinolytic activity and elevated plasminogen activator inhibitor-1. *Metabolism* **1990**, *39*, 1044–1048. [CrossRef]

91. Pasarica, M.; Gowronska-Kozak, B.; Burk, D.; Remedios, I.; Hymel, D.; Gimble, J.; Ravussin, E.; Bray, G.A.; Smith, S.R. Adipose tissue collagen VI in obesity. *J. Clin. Endocrinol. Metab.* **2009**, *94*, 5155–5162. [CrossRef] [PubMed]

92. Tartare-Deckert, S. Matrix metalloproteinases are differentially expressed in, adipose tissue during obesity and modulate adipocyte differentiation. *J. Biol. Chem.* **2003**, *278*, 11888–11896.

93. Franco, F.O.; Lopes, M.A.; Henriques, F.S.; Neves, R.X.; Bianchi Filho, C.; Batista, M.L., Jr. Cancer cachexia differentially regulates visceral adipose tissue turnover. *J. Endocrinol.* **2017**, *232*, 493–500. [CrossRef]

94. Buechler, C.; Krautbauer, S.; Eisinger, K. Adipose tissue fibrosis. *World J. Diabetes* **2015**, *6*, 548–553. [CrossRef]

95. Dossus, L.; Jimenez-Corona, A.; Romieu, I.; Boutron-Ruault, M.C.; Boutten, A.; Dupré, T.; Fagherazzi, G.; Clavel-Chapelon, F.; Mesrine, S. C-reactive protein and postmenopausal breast cancer risk: Results from the E3N cohort study. *Cancer Causes Control.* **2014**, *25*, 533–539. [CrossRef]

96. Engin, A.B. Adipocyte-macrophage cross-talk in obesity. *Adv. Exp. Med. Biol.* **2017**, *960*, 327–343.

97. Koru-Sengul, T.; Santander, A.M.; Miao, F.; Sanchez, L.G.; Jorda, M.; Glück, S.; Ince, T.A.; Nadji, M.; Chen, Z.; Penichet, M.L.; et al. Breast cancers from black women exhibit higher numbers of immunosuppressive macrophages with proliferative activity and of crown-like structures associated with lower survival compared to non-black latinas and caucasians. *Breast Cancer Res. Treat.* **2016**, *158*, 113–126. [CrossRef]

98. Gilbert, C.A.; Slingerland, J.M. Cytokines, obesity, and cancer: New insights on mechanisms linking obesity to cancer risk and progression. *Annu. Rev. Med.* **2013**, *64*, 45–57. [CrossRef]

99. Berger, N.A. Crown-like Structures in Breast Adipose Tissue from Normal Weight Women: Important Impact. *Cancer Prev. Res.* **2017**, *10*, 223–225. [CrossRef]

100. Iyengar, N.M.; Brown, K.A.; Zhou, X.K.; Gucalp, A.A.; Subbaramaiah, K.; Giri, D.D.; Zahid, H.; Bhardwaj, P.; Wendel, N.K.; Falcone, D.J.; et al. Metabolic obesity, adipose inflammation and elevated breast aromatase in women with normal body mass index. *Cancer Prev. Res. (Phila)* **2017**, *10*, 235–243. [CrossRef]

101. Iyengar, N.M.; Zhou, X.K.; Gucalp, A.; Morris, P.G.; Howe, L.R.; Giri, D.D.; Morrow, M.; Wang, H.; Pollak, M.; Jones, L.W.; et al. Systemic correlates of white adipose tissue inflammation in early-stage Breast Cancer. *Clin. Cancer Res.* **2016**, *22*, 2283–2289. [CrossRef] [PubMed]

102. Gyamfi, J.; Eom, M.; Koo, J.S.; Choi, J. Multifaceted Roles of Interleukin-6 in Adipocyte-Breast Cancer Cell Interaction. *Transl. Oncol.* **2018**, *11*, 275–285. [CrossRef] [PubMed]

103. Knupfer, H.; Preiss, R. Significance of interleukin-6 (IL-6) in breast cancer (review). *Breast Cancer Res. Treat.* **2007**, *102*, 129–135. [CrossRef] [PubMed]

104. Alfano, C.M.; Peng, J.; Andridge, R.R.; Lindgren, M.E.; Povoski, S.P.; Lipari, A.M.; Agnese, D.M.; Farrar, W.B.; Yee, L.D.; Carson, W.E.; et al. Inflammatory cytokines and comorbidity development in Breast Cancer survivors versus noncancer controls: Evidence for accelerated aging? *J. Clin. Oncol.* **2017**, *35*, 149–156. [CrossRef]

105. Balkwill, F. Tumour necrosis factor and cancer. *Nat. Rev. Cancer* **2009**, *9*, 361–371. [CrossRef]

106. Liu, D.; Wang, X.; Chen, Z. Tumor necrosis factor-alpha, a regulator and therapeutic agent on Breast Cancer. *Curr. Pharm. Biotechnol.* **2016**, *17*, 486–494. [CrossRef]

107. Zhang, H.H.; Halbleib, M.; Ahmad, F.; Manganiello, V.C.; Greenberg, A.S. Tumor necrosis factor-alpha stimulates lipolysis in differentiated human adipocytes through activation of extracellular signal-related kinase and elevation of intracellular cAMP. *Diabetes* **2002**, *51*, 2929–2935. [CrossRef]

108. Péqueux, C.; Raymond-Letron, I.; Blacher, S.; Boudou, F.; Adlanmerini, M.; Fouque, M.J.; Rochaix, P.; Noel, A.; Foidart, J.M.; Krust, A.; et al. Stromal estrogen receptor promotes tumor growth by normalizing an increased angiogenesis. *Cancer Res.* **2012**, *72*, 3010–3019. [CrossRef]

109. Grodin, J.M.; Siiteri, P.K.; MacDonald, P.C. Source of estrogen production in postmenopausal women. *J. Clin. Endocrinol. Metab.* **1973**, *36*, 207–214. [CrossRef]

110. van Landeghem, A.A.; Poortman, J.; Nabuurs, M.; Thijssen, J.H. Endogenous concentration and subcellular distribution of estrogens in normal and malignant human breast tissue. *Cancer Res.* **1985**, *45*, 2900–2906.

111. Rubin, G.L.; Zhao, Y.; Kalus, A.M.; Simpson, E.R. Peroxisome proliferator-activated receptor gamma ligands inhibit estrogen biosynthesis in human breast adipose tissue: Possible implications for breast cancer therapy. *Cancer Res.* **2000**, *60*, 1604–1608. [PubMed]

112. Dalamaga, M.; Diakopoulos, K.N.; Mantzoros, C.S. The role of adiponectin in cancer: A review of current evidence. *Endocr. Rev.* **2012**, *33*, 547–594. [CrossRef]

113. Kang, J.H.; Lee, Y.Y.; Yu, B.Y.; Yang, B.S.; Cho, K.H.; Yoon, D.K.; Roh, Y.K. Adiponectin induces growth arrest and apoptosis of MDA-MB-231 breast cancer cell. *Arch. Pharm. Res.* **2005**, *28*, 1263–1269. [CrossRef] [PubMed]

114. Chung, S.J.; Nagaraju, G.P.; Nagalingam, A.; Muniraj, N.; Kuppusamy, P.; Walker, A.; Woo, J.; Gyrffy, B.; Gabrielson, E.; Saxena, N.K.; et al. ADIPOQ/adiponectin induces cytotoxic autophagy in breast cancer cells through STK11/LKB1-mediated activation of the AMPK-ULK1 axis. *Autophagy* **2017**, *13*, 1386–1403. [CrossRef] [PubMed]

115. Panno, M.L.; Naimo, G.D.; Spina, E.; Andò, S.; Mauro, L. Different molecular signaling sustaining adiponectin action in breast cancer. *Curr. Opin. Pharmacol.* **2016**, *31*, 1–7. [CrossRef] [PubMed]

116. Garofalo, C.; Koda, M.; Cascio, S.; Sulkowska, M.; Kanczuga-Koda, L.; Golaszewska, J.; Russo, A.; Sulkowski, S.; Surmacz, E. Increased expression of leptin and the leptin receptor as a marker of breast cancer progression: Possible role of obesity-related stimuli. *Clin. Cancer Res.* **2006**, *12*, 1447–1453. [CrossRef] [PubMed]

117. Garofalo, C.; Surmacz, E. Leptin and cancer. *J. Cell Physiol.* **2006**, *207*, 12–22. [CrossRef]

118. Feldman, D.E.; Chen, C.; Punj, V.; Tsukamoto, H.; Machida, K. Pluripotency factor-mediated expression of the leptin receptor (OB-R) links obesity to oncogenesis through tumor-initiating stem cells. *Proc. Natl. Acad. Sci. USA* **2012**, *109*, 829–834. [CrossRef]

119. Fusco, R.; Galgani, M.; Procaccini, C.; Franco, R.; Pirozzi, G.; Fucci, L.; Laccetti, P.; Matarese, G. Cellular and molecular crosstalk between leptin receptor and estrogen receptor-{alpha} in breast cancer: Molecular basis for a novel therapeutic setting. *Endocr. Relat. Cancer* **2010**, *17*, 373–382. [CrossRef]

120. Strong, A.L.; Ohlstein, J.F.; Biagas, B.A.; Rhodes, L.V.; Pei, D.T.; Tucker, H.A.; Llamas, C.; Bowles, A.C.; Dutreil, M.F.; Zhang, S.; et al. Leptin produced by obese adipose stromal/stem cells enhances proliferation and metastasis of estrogen receptor positive breast cancers. *Breast Cancer Res.* **2015**, *17*, 112. [CrossRef]

121. Zhang, C.; Yue, C.; Herrmann, A.; Song, J.; Egelston, C.; Wang, T.; Zhang, Z.; Li, W.; Lee, H.; Aftabizadeh, M.; et al. STAT3 Activation-Induced Fatty Acid Oxidation in CD8(+) T Effector Cells Is Critical for Obesity-Promoted Breast Tumor Growth. *Cell Metab.* **2020**, *31*, 148–161. [CrossRef] [PubMed]

122. Wang, Z.; Aguilar, E.G.; Luna, J.I.; Dunai, C.; Khuat, L.T.; Le, C.T.; Mirsoian, A.; Minnar, C.M.; Stoffel, K.M.; Sturgill, I.R.; et al. Paradoxical effects of obesity on T cell function during tumor progression and PD-1 checkpoint blockade. *Nat. Med.* **2019**, *25*, 141–151. [CrossRef] [PubMed]

123. Douros, J.D.; Baltzegar, D.A.; Reading, B.J.; Seale, A.P.; Lerner, D.T.; Grau, E.G.; Borski, R.J. Leptin Stimulates Cellular Glycolysis Through a STAT3 Dependent Mechanism in Tilapia. *Front. Endocrinol. (Lausanne)* **2018**, *9*, 465. [CrossRef]

124. Ollberding, N.J.; Kim, Y.; Shvetsov, Y.B.; Wilkens, L.R.; Franke, A.A.; Cooney, R.V.; Maskarinec, G.; Hernandez, B.Y.; Henderson, B.E.; Le Marchand, L.; et al. Prediagnostic leptin, adiponectin, C-reactive protein, and the risk of postmenopausal breast cancer. *Cancer Prev. Res. (Phila)* **2013**, *6*, 188–195. [CrossRef]

125. Sultana, R.; Kataki, A.C.; Borthakur, B.B.; Basumatary, T.K.; Bose, S. Imbalance in leptin-adiponectin levels and leptin receptor expression as chief contributors to triple negative breast cancer progression in Northeast India. *Gene* **2017**, *621*, 51–58. [CrossRef]

126. Lee, D.; Suh, D.S.; Lee, S.C.; Tigyi, G.J.; Kim, J.H. Role of autotaxin in cancer stem cells. *Cancer Metastasis Rev.* **2018**, *37*, 509–518. [CrossRef]

127. Yang, S.Y.; Lee, J.; Park, C.G.; Kim, S.; Hong, S.; Chung, H.C.; Min, S.K.; Han, J.W.; Lee, H.W.; Lee, H.Y. Expression of autotaxin (NPP-2) is closely linked to invasiveness of breast cancer cells. *Clin. Exp. Metastasis* **2002**, *19*, 603–608. [CrossRef]

128. Steppan, C.M.; Bailey, S.T.; Bhat, S.; Brown, E.J.; Banerjee, R.R.; Wright, C.M.; Patel, H.R.; Ahima, R.S.; Lazar, M.A. The hormone resistin links obesity to diabetes. *Nature* **2001**, *409*, 307–312. [CrossRef]

129. Avtanski, D.; Garcia, A.; Caraballo, B.; Thangeswaran, P.; Marin, S.; Bianco, J.; Lavi, A.; Poretsky, L. Resistin induces breast cancer cells epithelial to mesenchymal transition (EMT) and stemness through both adenylyl cyclase-associated protein 1 (CAP1)-dependent and CAP1-independent mechanisms. *Cytokine* **2019**, *120*, 155–164. [CrossRef]

130. Rosendahl, A.H.; Bergqvist, M.; Lettiero, B.; Kimbung, S.; Borgquist, S. Adipocytes and obesity-related conditions jointly promote Breast Cancer cell growth and motility: Associations with CAP1 for prognosis. *Front. Endocrinol. (Lausanne)* **2018**, *9*, 689. [CrossRef]

131. Gallego, M.I.; Bierie, B.; Hennighausen, L. Targeted expression of HGF/SF in mouse mammary epithelium leads to metastatic adenosquamous carcinomas through the activation of multiple signal transduction pathways. *Oncogene* **2003**, *22*, 8498–8508. [CrossRef] [PubMed]

132. Edakuni, G.; Sasatomi, E.; Satoh, T.; Tokunaga, O.; Miyazaki, K. Expression of the hepatocyte growth factor/c-Met pathway is increased at the cancer front in breast carcinoma. *Pathol. Int.* **2001**, *51*, 172–178. [CrossRef] [PubMed]

133. Iyengar, P.; Espina, V.; Williams, T.W.; Lin, Y.; Berry, D.; Jelicks, L.A.; Lee, H.; Temple, K.; Graves, R.; Pollard, J.; et al. Adipocyte-derived collagen VI affects early mammary tumor progression in vivo, demonstrating a critical interaction in the tumor/stroma microenvironment. *J. Clin. Investig.* **2005**, *115*, 1163–1176. [CrossRef] [PubMed]

134. Park, J.; Scherer, P.E. Adipocyte-derived endotrophin promotes malignant tumor progression. *J. Clin. Investig.* **2012**, *22*, 4243–4256. [CrossRef]

135. Motrescu, E.R.; Rio, M.C. Cancer cells, adipocytes and matrix metalloproteinase 11: A vicious tumor progression cycle. *Biol. Chem.* **2008**, *389*, 1037–1041. [CrossRef]

136. Huang, J.; Duran, A.; Reina-Campos, M.; Valencia, T.; Castilla, E.A.; Muller, T.D.; Tschep, M.H.; Moscat, J.; Diaz-Meco, M.T. Adipocyte p62/SQSTM1 Suppresses Tumorigenesis through Opposite Regulations of Metabolism in Adipose Tissue and Tumor. *Cancer Cell* **2018**, *33*, 770–784. [CrossRef]

137. Pope, B.D.; Warren, C.R.; Parker, K.K.; Cowan, C.A. Microenvironmental Control of Adipocyte Fate and Function. *Trends Cell Biol.* **2016**, *26*, 745–755. [CrossRef]
138. Pascual, G.; Avgustinova, A.; Mejetta, S.; Martin, M.; Castellanos, A.; Attolini, C.S.; Berenguer, A.; Prats, N.; Toll, A.; Hueto, J.A.; et al. Targeting metastasis-initiating cells through the fatty acid receptor CD36. *Nature* **2017**, *541*, 41–45. [CrossRef]
139. Aznar Benitah, S. Metastatic-initiating cells and lipid metabolism. *Cell Stress* **2017**, *1*, 110–114. [CrossRef]
140. Nath, A.; Chan, C. Genetic alterations in fatty acid transport and metabolism genes are associated with metastatic progression and poor prognosis of human cancers. *Sci. Rep.* **2016**, *6*, 18669. [CrossRef]
141. Li, Z.; Wu, Q.; Sun, S.; Wu, J.; Li, J.; Zhang, Y.; Wang, C.; Yuan, J.; Sun, S. Monocarboxylate transporters in breast cancer and adipose tissue are novel biomarkers and potential therapeutic targets. *Biochem. Biophys. Res. Commun.* **2018**, *501*, 962–967. [CrossRef]
142. Wu, Q.; Li, J.; Li, Z.; Sun, S.; Zhu, S.; Wang, L.; Wu, J.; Yuan, J.; Zhang, Y.; Sun, S.; et al. Exosomes from-áthe tumour-adipocyte interplay stimulate beige/brown differentiation and reprogram metabolism in stromal adipocytes to promote tumour progression. *J. Exp. Clin. Cancer Res.* **2019**, *38*, 223. [CrossRef]
143. Martinez-Outschoorn, U.E.; Lin, Z.; Whitaker-Menezes, D.; Howell, A.; Lisanti, M.P.; Sotgia, F. Ketone bodies and two-compartment tumor metabolism: Stromal ketone production fuels mitochondrial biogenesis in epithelial cancer cells. *Cell Cycle* **2012**, *11*, 3956–3963. [CrossRef] [PubMed]
144. Huang, C.K.; Chang, P.H.; Kuo, W.H.; Chen, C.L.; Jeng, Y.M.; Chang, K.J.; Shew, J.Y.; Hu, C.M.; Lee, W.H. Adipocytes promote malignant growth of breast tumours with monocarboxylate transporter 2 expression via +¦-hydroxybutyrate. *Nat. Commun.* **2017**, *8*, 14706. [CrossRef] [PubMed]
145. Argilés, J.M.; Busquets, S.; Stemmler, B.; Lopez-Soriano, F.J. Cancer cachexia: Understanding the molecular basis. *Nat. Rev. Cancer* **2014**, *14*, 754–762. [CrossRef]
146. Fearon, K.C.H. Cancer cachexia: Developing multimodal therapy for a multidimensional problem. *Eur. J. Cancer* **2008**, *44*, 1124–1132. [CrossRef]
147. Mundy, G.R. Metastasis to bone: Causes, consequences and therapeutic opportunities. *Nat. Rev. Cancer* **2002**, *2*, 584–593. [CrossRef]
148. Neves, R.X.; Rosa-Neto, J.C.; Yamashita, A.S.; Matos-Neto, E.M.; Riccardi, D.M.R.; Lira, F.S.; Batista, M.L., Jr.; Seelaender, M. White adipose tissue cells and the progression of cachexia: Inflammatory pathways. *J. Cachexia Sarcopenia Muscle* **2016**, *7*, 193–203. [CrossRef]
149. Daas, S.I.; Rizeq, B.R.; Nasrallah, G.K. Adipose tissue dysfunction in cancer cachexia. *J. Cell Physiol.* **2018**, *234*, 13–22. [CrossRef]
150. COSTA, G.; Holland, J.F. Effects of Krebs-2 carcinoma on the lipide metabolism of male Swiss mice. *Cancer Res.* **1962**, *22*, 1081–1083.
151. Das, S.K.; Eder, S.; Schauer, S.; Diwoky, C.; Temmel, H.; Guertl, B.; Gorkiewicz, G.; Tamilarasan, K.P.; Kumari, P.; Trauner, M.; et al. Adipose triglyceride lipase contributes to cancer-associated cachexia. *Science* **2011**, *333*, 233–238. [CrossRef] [PubMed]
152. Petruzzelli, M.; Wagner, E.F. Mechanisms of metabolic dysfunction in cancer-associated cachexia. *Genes. Dev.* **2016**, *30*, 489–501. [CrossRef] [PubMed]
153. Bodine, S.C.; Baehr, L.M. Skeletal muscle atrophy and the E3 ubiquitin ligases MuRF1 and MAFbx/atrogin-1. *Am. J. Physiol. Endocrinol. Metab.* **2014**, *307*, E469–E484. [CrossRef] [PubMed]
154. Vaitkus, J.A.; Celi, F.S. The role of adipose tissue in cancer-associated cachexia. *Exp. Biol. Med. (Maywood)* **2017**, *242*, 473–481. [CrossRef] [PubMed]
155. Bing, C.; Trayhurn, P. Regulation of adipose tissue metabolism in cancer cachexia. *Curr. Opin. Clin. Nutr. Metab. Care* **2008**, *11*, 201–207. [CrossRef] [PubMed]
156. Shaw, J.H.; Wolfe, R.R. Fatty acid and glycerol kinetics in septic patients and in patients with gastrointestinal cancer. The response to glucose infusion and parenteral feeding. *Ann. Surg.* **1987**, *205*, 368–376. [CrossRef]
157. Das, S.K.; Hoefler, G. The role of triglyceride lipases in cancer associated cachexia. *Trends Mol. Med.* **2013**, *19*, 292–301. [CrossRef]
158. Boden, G. Obesity and free fatty acids. *Endocrinol. Metab. Clin. N. Am.* **2008**, *37*, 635–646. [CrossRef]
159. Zhou, Q.; Du, J.; Hu, Z.; Walsh, K.; Wang, X.H. Evidence for adipose-muscle cross talk: Opposing regulation of muscle proteolysis by adiponectin and Fatty acids. *Endocrinology* **2007**, *148*, 5696–5705. [CrossRef]
160. Baracos, V.E.; Arribas, L. Sarcopenic obesity: Hidden muscle wasting and its impact for survival and complications of cancer therapy. *Ann. Oncol.* **2018**, *29*, ii1–ii9. [CrossRef]

161. Valencia, T.; Kim, J.Y.; Abu-Baker, S.; Moscat-Pardos, J.; Ahn, C.S.; Reina-Campos, M.; Duran, A.; Castilla, E.A.; Metallo, C.M.; Disaz-Meco, M.T.; et al. Metabolic reprogramming of stromal fibroblasts through p62-mTORC1 signaling promotes inflammation and tumorigenesis. *Cancer Cell* **2014**, *26*, 121–135. [CrossRef] [PubMed]
162. Ishay-Ronen, D.; Diepenbruck, M.; Kalathur, R.K.R.; Sugiyama, N.; Tiede, S.; Ivanek, R.; Bantug, G.; Morini, M.F.; Wang, J.; Hess, C.; et al. Gain Fat-Lose Metastasis: Converting Invasive Breast Cancer Cells into Adipocytes Inhibits Cancer Metastasis. *Cancer Cell* **2019**, *35*, 17–32. [CrossRef] [PubMed]

Review

Esophageal Cancer Development: Crucial Clues Arising from the Extracellular Matrix

Antonio Palumbo Jr. [1,*], Nathalia Meireles Da Costa [2], Bruno Pontes [3], Felipe Leite de Oliveira [4], Matheus Lohan Codeço [1], Luis Felipe Ribeiro Pinto [2] and Luiz Eurico Nasciutti [1]

[1] Laboratório de Interações Celulares, Instituto de Ciências Biomédicas, Universidade Federal do Rio de Janeiro Prédio de Ciências da Saúde—Cidade Universitária, Ilha do Fundão, Av. Carlos Chagas, 373—bloco F, sala 26, Rio de Janeiro CEP: 21941-902, RJ, Brazil; matheuslohan@hotmail.com (M.L.C.); luiz.nasciutti@histo.ufrj.br (L.E.N.)

[2] Programa de Carcinogênese Molecular, Instituto Nacional de Câncer—INCA, Rua André Cavalcanti, 37—Centro, Rio de Janeiro CEP: 20231-050, RJ, Brazil; nathalia.meireles@inca.gov.br (N.M.D.C.); lfrpinto@inca.gov.br (L.F.R.P.)

[3] Laboratório de Pinças Óticas (LPO-COPEA), Instituto de Ciências Biomédicas, Universidade Federal do Rio de Janeiro, Rio de Janeiro 21941-902, RJ, Brazil; brunoaccpontes@gmail.com

[4] Laboratório de Proliferação e Diferenciação Celular, Instituto de Ciências Biomédicas (ICB), Universidade Federal do Rio de Janeiro, Rio de Janeiro 21941-902, RJ, Brazil; felipe@histo.ufrj.br

* Correspondence: palumbo@icb.ufrj.br; Tel.: +55-21-39386476

Received: 30 December 2019; Accepted: 13 February 2020; Published: 17 February 2020

Abstract: In the last years, the extracellular matrix (ECM) has been reported as playing a relevant role in esophageal cancer (EC) development, with this compartment being related to several aspects of EC genesis and progression. This sounds very interesting due to the complexity of this highly incident and lethal tumor, which takes the sixth position in mortality among all tumor types worldwide. The well-established increase in ECM stiffness, which is able to trigger mechanotransduction signaling, is capable of regulating several malignant behaviors by converting alteration in ECM mechanics into cytoplasmatic biochemical signals. In this sense, it has been shown that some molecules play a key role in these events, particularly the different collagen isoforms, as well as enzymes related to its turnover, such as lysyl oxidase (LOX) and matrix metalloproteinases (MMPs). In fact, MMPs are not only involved in ECM stiffness, but also in other events related to ECM homeostasis, which includes ECM remodeling. Therefore, the crucial role of distinct MMPs isoform has already been reported, especially MMP-2, -3, -7, and -9, along EC development, thus strongly associating these proteins with the control of important cellular events during tumor progression, particularly in the process of invasion during metastasis establishment. In addition, by distinct mechanisms, a vast diversity of glycoproteins and proteoglycans, such as laminin, fibronectin, tenascin C, galectin, dermatan sulfate, and hyaluronic acid exert remarkable effects in esophageal malignant cells due to the activation of oncogenic signaling pathways mainly involved in cytoskeleton alterations during adhesion and migration processes. Finally, the wide spectrum of interactions potentially mediated by ECM may represent a singular intervention scenario in esophageal carcinogenesis natural history and, due to the scarce knowledge on the cellular and molecular mechanisms involved in EC development, the growing body of evidence on ECM's role along esophageal carcinogenesis might provide a solid base to improve its management in the future.

Keywords: esophageal carcinogenesis; extracellular matrix; stiffness; remodeling; adhesion; metalloproteinases; glycoproteins; proteoglycans

---

1. Introduction

The high level of organization among multicellular organisms is largely attributed to a constant process of communication between different cellular types [1,2]. In fact, due to this highly organized communication system, a myriad of chemical signals is stratified, thus integrating physically distant compartments in a functional organism [3,4]. However, the physical and chemical integration is not only based on the establishment of cell–cell communications, but also on the communication between cells and their surrounding microenvironment, particularly the extracellular matrix (ECM) [5,6]. ECM can be defined as a complex non-cellular compartment, present within tissues and organs, providing physical scaffolding, as well as biochemical and biomechanical signals required for several biological processes [1]. In this sense, the establishment of interactions between cells and ECM is crucial for morphogenesis and homeostasis maintenance, besides presenting an important role in the genesis and progression of distinct diseases, including cancer [7,8]. Thus, the relevance of ECM in cancer has strongly increased over the last years, and it is now well demonstrated that it does not represent only a mere physical compartment, but a functional protein network, capable of modulating the fate of key events during carcinogenesis, such as cell survival, apoptosis, proliferation, angiogenesis, migration, and invasion, among others [7–12]. Additionally, several biological aspects of ECM (adhesion, stiffness, remodeling, growth factor release, and others) may cooperate and stimulate the malignant behavior of cancer cells by a feedback mechanism that is established between tumors and their surrounding microenvironment [13,14]. Moreover, in the last years, it has been reported that ECM plays a relevant role in esophageal cancer (EC) development, being this compartment related to several aspects of its genesis and progression [15]. EC is a highly incident and lethal cancer, taking the sixth position in mortality in men, among all tumor types worldwide [16]. This lethal tumor confers a 5-year survival rate of about 15–25% of patients, demonstrating its poor prognosis [17]. There are two main EC histopathological subtypes: esophageal squamous cell carcinoma (ESCC) and esophageal adenocarcinoma (EAC), which widely differ considering populations affected, etiological factors, and molecular alterations, among others. Although ESCC development is highly associated with tobacco and alcohol abuse and ingestion of high temperature beverages, for EAC, the main associated risk factors are obesity, gastroesophageal reflux disease (GERD), and Barrett's esophagus (BE), an intestinal metaplasia where the normal stratified squamous esophageal epithelium is replaced by a columnar intestinal-like one [18,19]. ESCC represents the predominant EC histotype; nevertheless, along with the increase in obesity rates in some western countries, the incidence of EAC has increased sharply over the past few decades [20,21]. Although EC is a remarkably incident and lethal cancer, the knowledge on its biology is still scarce. EC poor prognosis is directly associated with its late detection, which is a consequence of the lack of clinical symptoms in early tumor stages. Thus, the deeper understanding of the mechanisms involved in its genesis and/or progression may be useful in identifying potential markers for diagnosis and prognosis, as well as potential therapeutic targets. Therefore, the reciprocal interactions that occur between the cells and the surrounding ECM orchestrate a complex cascade of events during esophageal malignant transformation, where the adhesion mediated by glycoproteins and proteoglycans triggers signaling pathways, which induce the expression and activation of catalytic enzymes that, in turn, promote structural alterations in ECM stiffness and remodeling [15]. These alterations culminate in the activation of signaling pathways in a feedback loop mechanism (Figure 1). In this context, the interactions between EC and ECM may shed some light on the cellular and molecular mechanisms that govern the malignant development of these tumors. Thus, the aim of this review was to compile existing data of crucial phenomena related to ECM, focusing specifically on the stiffness, remodeling, and adhesion/migration events under the context of esophageal carcinogenesis, as presented in the following sections.

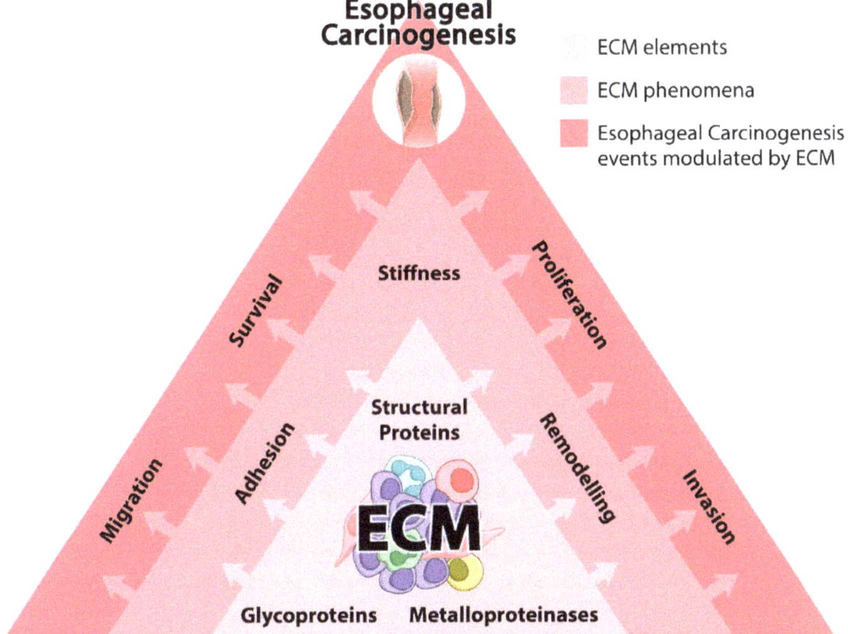

Figure 1. Extracellular matrix (ECM) and esophageal carcinogenesis. The figure schematically represents how ECM may affect the genesis and/or development of esophageal tumors. The first layer represents the main classes of ECM constituting proteins involved in esophageal carcinogenesis. The second layer represents ECM phenomena modulated in esophageal tissues as a consequence of the altered protein constitution and function of this compartment. The third layer represents the crucial cellular events impacted by structural and functional alterations occurring in ECM, which may play a crucial role in esophageal carcinogenesis. Arrows indicate the dynamism and hierarchy between the three layers of events.

2. EC and ECM

2.1. Structural Proteins and ECM Stiffness

The tumor microenvironment plays a key role in tumor development and progression [15,22–26]. This dynamic environment is formed by several cellular components that include not only cancer cells, but also fibroblasts, and immune and endothelial cells [15]. Moreover, one of the most important constituents of the tumor microenvironment is the ECM, a meshwork of proteins and glycosaminoglycans surrounding tumor cells [12]. ECM provides the biochemical and mechanical support for tumor progression [12,27,28]. The expression of several ECM proteins has been shown to be upregulated in tumors, of which, the most abundant is type I collagen [29]. This structural protein not only fills up the interstitial spaces between cancer cells, but it also changes the mechanical properties of cancer tissues, particularly by increasing their stiffness and providing tissues with tensile strength and resistance to deformation [12,30]. The increase in tissue stiffness occurs not only due to the increased deposition and cross-linking of thick fibers of type I collagen together with other proteins, such as fibronectin and types III and IV collagens, but also by the increase in cross-linking capacity of these meshworks [29]. These modifications involve several post-translational changes in the molecule that take place inside the cells, followed by reactions that occur in the ECM [31]. For instance, lysine or hydroxylysine residues in collagen molecule can undergo oxidative deamination catalyzed by lysyl oxidase (LOX) enzymes at the ECM, leading to the formation of covalent intermolecular cross-links between neighboring collagen molecules [31]. Although type I collagen was traditionally considered as a physical barrier to prevent tumor cell dissemination, mainly because early studies in the field showed that collagen molecules should be degraded to allow tumor progression, it is now evident that type I collagen is actively involved in tumor invasion [32]. It is well-established that the increase in ECM structure and stiffness is directly associated with tumor progression. The so-called "tumor stiffening" is due to the deposition and cross-linking of several ECM molecules [29,33–37]. Huge amounts of ECM molecules are deposited during EC development [38]. Thus, the constituents of the ECM can have a lasting effect on cancer cells. Indeed, ECM remodeling creates a reorganized environment that promotes tumor progression by destabilizing cell polarity and cell–cell adhesions and by increasing growth factor signaling [32], which impact on tumor gene expression, differentiation, proliferation, migration, and responses to treatments [32]. Type I collagen protein secretion within the tumor microenvironment is mainly originated from the cells surrounding these tumors, particularly cancer-associated fibroblasts (CAFs) or tumor-associated fibroblasts (TAFs) [39]. These cells have a specially activated phenotype, marked by the expression of fibroblast activation protein-α (FAP-α) and α-smooth muscle actin (α-SMA), although with extremely heterogeneous subpopulations that may have different roles during tumor development and progression [40,41]. The increase in expression of these CAF proteins are induced by growth factors and microRNAs (miRNAs) secreted by cancer cells [22,42,43]. Once in this state, CAFs can modulate the tumor progression in several ways, such as via secreted factors, activating pro-inflammatory pathways, disrupting immune surveillance, and also by altering the ECM protein structure and stiffness [22,44–48]. Regarding EC, a recent study demonstrated that cells from tumor microenvironment synthesize increased ECM proteins, and the upregulation of these proteins is associated with patients' poor prognosis and chemoresistance [38]. Moreover, using cell-derived 3D ECM, the authors showed that differences in ECM composition and structure are both crucial to EC cell response to chemotherapeutic drugs [38]. Thus, targeting ECM proteins synthesized by tumor surrounding cells could be an effective method to control cancer development and progression mainly because increased levels of ECM proteins together with their structural organization both act as a barrier for chemotherapy efficiency. A combinatory therapy that acts not only in EC cells but also in the ECM surrounding these cells should be considered. Tumor cells also need to degrade and remodel type I collagen meshwork. This is mainly performed by matrix metalloproteinases (MMPs) secreted by tumor cells as a very important feature that allows for the invasion and dissemination of cancers. Collagen molecules are synthesized as procollagens by the cell. When procollagen is secreted extracellularly, it is cleaved, assembled, and cross-linked to form collagen [12]. Indeed, the levels of such propeptides are associated with worse prognosis of patients affected by different types of cancers, being detected in their serum [49]. Moreover, it was already demonstrated that collagen fragments could induce the expression of signaling molecules, such as vascular endothelial growth factor (VEGF) and C-X-C motif chemokine receptor 4 (CXCR4), known as activators of MMP-2 and MMP-9 [50]. Over the past few decades, the notion that tumors secrete MMPs to degrade type I collagen in ECM has been very well accepted [51]. Recent studies, however, have described how cancer cells also produce their own type I collagen molecules, nevertheless presenting a different molecular weight from that of stromal cells, particularly in the case of EC [52,53]. Cancer-derived type I collagen molecules may affect tumor microenvironment and, although the precise mechanisms need to be explored, it is already known that tumor type I collagen facilitates cancer cell adhesion by modulating ECM to contribute to tumor clone formation [52,53]. Thus, one could suggest that cancer-derived type I collagen represents a promising target to improve cancer diagnosis and treatment, pointing out the need to further elucidate their function, particularly in EC. Apart from ECM remodeling promoted by tumor cells, there is also little research trying to elucidate the effect of CAFs on EC ECM remodeling. By observing the morphology of collagen fibers using multi-photon laser scanning microscopy with the second-harmonic generation, Hanley and colleagues showed that the appearance of elongated collagen fibers within CAFs is associated with poor survival rates in EC patients [54]. Moreover, the authors also showed that the formation of this collagen linear structures are specifically regulated by a CAF type that displays a myofibroblastic phenotype (αSMA-positive), suggesting that these αSMA-positive CAFs are the ones capable of regulating collagen fiber elongation and highlighting that the combined evaluation of collagen structure and αSMA-positive CAFs may be an important feature for the prognostic assessment of EC [54]. Cells sense ECM stiffness through their transmembrane receptors that are linked to the actomyosin cytoskeleton. Thus, the mechanical properties of ECM can influence growth, survival, migration, invasion, and differentiation of cells within their tissues [12,30]. Once the development of pathologic conditions, such as cancer, are accompanied by alterations in the composition, organization, and mechanical properties of ECM, these changes contribute not only to the malignant transformation but also to tumor progression and metastasis [12,30]. Cells bind to the ECM through diverse transmembrane receptors, such as integrins, syndecans, and, particularly in the case of type I collagen molecules, to discoidin domain receptors (DDRs) [55]. Upon collagen binding, DDR receptors' cytoplasmic tails undergo autophosphorylation, creating docking sites for several other molecules, such as phosphoinositide 3-kinase (PI3K), extracellular-signal-regulated kinase (Erk), and myosin II [56–58]. This myosin II, together with actin, constantly checks the mechanical properties of ECM and propagates intracellular signaling pathways through a process known as mechanotransduction. It sounds very interesting, once it has already been reported that LOX, responsible for catalyzing the cross-links between collagen molecules, is highly expressed in ESCC samples, resulting in an increase in ECM stiffness [59], which, in turn, activates signaling pathways involved with malignant phenotype acquisition [12,30]. In fact, it was observed that the activation of specific signaling pathways in tumors triggered by alterations in ECM stiffness, besides altering cell morphology and migration [60]. Therefore, these data indicate that the stiffness patterns observed in EC samples [54,61] may represent the activation of several pathways involved in EC evolution. According to this hypothesis, the increase in ECM stiffness, which is observed in a vast range of tumor types [61], also contributes to tumor cells' sustained growth, by inducing telomerase activity and, consequently, telomere stretching [62]. Such a phenomenon was also observed along EC development [63]. In this scenario, as a consequence of telomere elongation, cells are unable to arrest along cell cycle progression, conferring a limitless replication pattern, which characterizes neoplastic cells [39]. Moreover, a previous study showed that the increase in integrin expression in epithelial cells is associated with a higher telomerase enzymatic activation [64]. Thus, once integrins represent the main class of receptors interacting with ECM, one could suggest that the activity of telomerase induced by the alterations in ECM stiffness, may be mediated, at least in part, by integrin receptors. On the other hand, it has been reported that loss of integrin subunits α6, β1, and β4 is related to ESCC patients' poor prognosis, whereas their presence is associated with increased relapse-free survival [65]. Thus, these data could suggest that the association between ECM stiffness—through integrin activation—and telomerase activity may represent an early step along esophageal carcinogenesis, once, during EC development, integrins are precociously lost [65]. In fact, it has been reported that telomere elongation is an early event related to esophageal carcinogenesis, as its association with esophageal pre-malignant lesions, such as BE, telomere elongation, and the consequently genetic instability triggered by this event, is also observed after the exposure to ethiopathological risk factors associated with EC development [63]. Finally, because integrins represent a key element during mechanotransduction, one could say that the alterations in ECM stiffness might play a central role in the esophageal tissue transformation due to its involvement in telomere elongation. In addition, a stiffer ECM will foster these signaling pathways, whereas a compliant ECM will halt them [12]. Thus, ECM stiffening through collagen cross-linking stimulates tumor cells in generating higher intracellular cytoskeletal tension, and also stronger traction forces on ECM [12]. ECM stiffening is positively correlated with the aggressive behavior of cancer cells, such that most cells migrate faster on stiffer substrates [66]. This feature can be explained because a stiffer ECM promotes cell migration and invasion by enhancing adhesion molecule clustering, focal adhesion kinase (FAK) phosphorylation, and Rho-GTPase activation, which favor the assembly of cellular protrusions such as filopodia and lamellipodia required for efficient tumor cell migration through ECM [67]. Moreover, cellular responses to mechanical stimuli also alter gene expression. For example, the expression of integrins, MMPs, and also cytoskeletal proteins are all increased in response to stiff substrates in a sort of positive feedback that enhances tumor progression and metastasis [68–70]. Although most of the mechanical mechanisms described above are already partially elucidated for some cancers, very little is known about these phenomena in EC. Therefore, investigating these characteristics in EC may be an interesting research topic for the field.

2.2. MMPs and ECM Remodeling

ECM remodeling is a very dynamic process, represented by the balance between production and degradation of proteins comprising ECM. This event plays an essential role in the maintenance of tissue homeostasis through the regulation of a myriad of interactions that constantly occur between cells and their surrounding microenvironment [71]. Nonetheless, the proteolytic enzymes play a central role in this process, by catalyzing the reactions that culminate in alterations in the chemical structure of target proteins [72]. In this context, the MMP protein family is characterized as multi-domain zinc-dependent endopeptidases and represent one of the major class of enzymes involved in ECM remodeling [72]. The different MMP subtypes are closely related to several distinct aspects of cancer development. Specifically, it has been reported that MMPs regulate key events in tumor progression, such as cell survival and invasion, metastasis development and angiogenesis, and malignant transformation [73]. MMPs are also biologically relevant for esophageal carcinogenesis. It has been previously shown that MMPs function as potential biomarkers, and distinct MMP subtypes, including MMP-1, MMP-2, MMP-3, MMP-7, and MMP-9, have been associated with EC diagnosis and/or prognosis [74–79]. In this way, among the nearly 30 distinct MMPs subtypes, MMP-2 and MMP-9 are the most strongly associated with esophageal carcinogenesis [74]. In fact, due to the breakdown of basement membrane collagen IV, MMP-2 and MMP-9 overexpression confers a poor prognosis for EC patients and is associated with late tumor stage, as well as with local invasion and presence of metastasis [74]. Hence, because aberrant expression of MMP-2 and MMP-9 plays a key role along EC progression, different studies have shed some light on the mechanisms involved in the deregulation of MMP-2 and MMP-9 expression. Accordingly, the study published by Wang and colleagues revealed that overexpression of aurora A kinases (AURKA) in ESCC cell lines triggers a signaling pathway involving other kinases, such as p38 mitogen-activated protein kinases (p38) and protein kinase B (Akt), that culminates in MMP-2 expression and activity augmentation and, consequently, enhancement of ESCC cells' invasion abilities [80]. The involvement of the mitogen-activated protein kinase (MAPK) pathway seems to be a central mechanism in the regulation of MMP-2 and MMP-9 expression, as it it was demonstrated that a decrease in extracellular signal-regulated kinases (Erk) 1/2 and p38 phosphorylation, promoted by the tumor suppressor distinct subgroup of the Ras family member 1 (*DIRAS1*), leads to a down regulation of both MMPs [75]. Moreover, in accordance with these data, the activation of c-Jun N-terminal kinase (JNK) by E3 ubiquitin-protein ligase RAD18 also promotes an upregulation of MMP-2 and MMP-9 expression, as well as an increment in ESCC malignant cell invasion capacity [76]. In addition, the involvement of MAPK members in MMP-2 and MMP-9 expression control is likely to be also indirectly regulated by DNA damage events, because DNA polymerase iota, which acts in translesion DNA repair, is able to increase the expression of both MMPs by inducing the activation of the JNK/activator protein 1 (AP-1) axis in ESCC cells [77]. Oppositely, the study performed by Garalla and colleagues showed that the blockage of the MAPK pathway was not able to interfere in MMP-7 expression [78], although MMP-7 aberrant levels had been previously related to EC progression [79,81]. Nonetheless, in the same study, the authors demonstrated that PI3K blockage, by using the chemical inhibitor LY294002, was able to strongly abolish MMP-7 secretion by EAC cells [78]. Additionally, the increase in MMP-7 levels could also be triggered through the modulation of other pathways, and, by using a panel of ESCC malignant cell lines, it was shown that the upregulation of AP-1 induced by activin A culminated with an increase in MMP-7 levels [82]. Taken together, these data suggest that different pathways are strictly associated with the modulation of distinct MMP subtypes' expression along esophageal carcinogenesis. Therefore, the well-established association between DNA damage and environmental factors, which characterizes the natural history of EC, may also be involved in the early deregulation of MMP function and expression. Furthermore, it seems that not only genetic alterations are related to the imbalance of MMP levels, but also to epigenetic mechanisms, such as acetylation of histones and DNA methylation, that could indirectly regulate MMP-2 and MMP-9 expression [83,84]. Additionally, obesity figures act as a crucial ethiopathological factor for EAC development. Corroborating this statement, high leptin levels, instead of adiponectin, are detected (among other hormonal alterations) in EAC patients. In this way, it was previously shown that the activation of signal transducer and activator of transcription 3 (STAT3) by leptin signaling pathway in the malignant EAC cell line OE-33 was capable of modulating the expression of MMP-2, but not that of MMP-9, which was also regulated by leptin, nevertheless in a STAT3-independent manner [85]. Because STAT3 is a recognized p38 activator, it could be reasonable to suggest that MAPK pathway activation by leptin may not be strong enough to regulate MMP-9 expression, or, as postulated by the authors, MMP-9 expression could be regulated by other mechanisms, including MAPK activators, or even other pathways mediated by leptin signaling. However, in the same work, Beales and colleagues demonstrated that adiponectin was able to antagonize leptin effects observed in EAC malignant cells, including MMP-2 and MMP-9 downregulation, thus indicating the importance of hormonal imbalance induced by obesity during EAC genesis and progression. In accordance with these observations, it was previously reported that obesity strongly influences the expression of MMP-2 and MMP-9, due to the fact that EAC cells display a great increase in the expression of these two enzymes after its co-cultivation with adipose tissue obtained from visceral fat area [86]. On the other hand, it was shown that the tumor suppressor candidate-activating transcription factor 3 (ATF3), which is significantly downregulated in malignant tissues [87], including ESCC specimens [88], coordinates a signaling cascade involved in the formation of a protein complex composed by p53, MDM2, and MMP-2, which mediates MMP-2 degradation and attenuation of tumor progression through cellular invasion inhibition [88]. Furthermore, in the same study, it was observed that cisplatin, the main chemotherapeutic agent used in EC management, was able to revert the invasive phenotype of ESCC cells, due to the induction of ATF3 expression occurring through p53 intermediation. These results sound very interesting because cisplatin is an alkylating agent that promotes DNA damage by forming adducts [89]. Otherwise, differently from other chemotherapeutic agents, cisplatin-based treatment may exhibit the best responses in EC patients, considering the additional effects exerted by this drug on the pathway that leads to MMP-2 degradation. In fact, MMP expression and activation seems to be regulated by different proteins that would be directly associated in a process orchestrated by the biological context. BE consists in one of the major hallmarks of EAC development, with GERD being the main inductor of this condition [90]. In this context, it has been previously reported that MMP-2 is activated by GERD occurring along EAC development [91]. In this way, chronic inflammation represents one of the main biological consequences promoted by GERD, with this inflammatory condition being a major driving force along EAC carcinogenesis [92]. Furthermore, nuclear factor kappa B (NF-kB), a key element activated during inflammation process, has been associated with the presence of invasion and metastasis during ESCC progression, by inducing epithelial mesenchymal transition (EMT) and MMP-9 expression [93]. Further, the study published by Shin and colleagues elucidated the mechanistic pathway linking the increase in MMP-9 expression and NF-kB by showing that protein tyrosine kinase 7 (PTK7) is able to activate NF-kB by a complex circuit, which involves different types of tyrosine kinase receptors, MAPK proteins, and Src [94]. In addition, it was previously reported that the pro-inflammatory interleukin 17 (IL-17) was able to induce the invasiveness of EAC cells by upregulating MMP-2 and MMP-9, and moreover increasing reactive oxygen species' (ROS) intracellular levels [95]. In fact, Lu and colleagues showed that high levels of oxidative stress, which characterizes the bile acid reflux environment, promotes apurinic/apyrimidinic endonuclease 1 (APE1) accumulation that, in addition to the promotion of cell survival through oxidative DNA damage repair and activation of STAT3 signaling, is also capable of binding to ADP-ribosylation factor 6 (ARF6) protein. The binding of APE1 to ARF6 leads to its stabilization and, subsequently, it introduces a recycling membrane-type 1 matrix metalloproteinase (MT1-MMP) to the cellular membrane, where the MT1-MMP enzyme will activate MMP-2 [91]. Moreover, along EAC evolution, the increase in nitric oxide levels, induced by GERD, was able to promote the invasion of high-grade dysplasia Barrett's cells by upregulating MMP-1 [96]. Of note, MMP-1 expression deregulation has been reported as a poor prognosis marker for EC development [97,98]. Therefore, it seems that a permissive environment would arise in esophageal tissue as a response to bile acid reflux insult. Specifically, the increase in NF-kB activation, as well as in ROS levels, caused by chronic inflammation plays a central role in the acquisition of a more invasive phenotype, as a consequence of the aberrant expression of MMPs, particularly MMP-2 and MMP-9. In addition, the relationship between GERD and MMPs along esophageal carcinogenesis seems to represent a wide event, and it has been reported that GERD condition is associated with the prevalence of polymorphisms in *MMP-1* (*1G/2G) and *MMP-3* (*6A/5A) [99]. Thus, because it is already known that polymorphisms in *MMP-1* (*1G/2G) and *MMP-3* (*6A/5A) are related to an enhanced risk for EAC development [100], these data suggest that the association between GERD and MMP polymorphisms is an early event during EAC development. Nevertheless, this scenario seems to be more complex—the impact of *MMP* polymorphisms on EC development risk modulation depends on the polymorphism itself, as well as on the *MMP* gene affected. In this way, a meta-analysis study conducted by Peng and colleagues revealed that the distinct polymorphisms present in *MMP-7* and *MMP-9* genes were not related to increased risk for EC development, and moreover, two polymorphisms found in *MMP-2* gene were associated with a diminished susceptibility of EC development [101]. Finally, it has been shown that epidermal growth factor (EGF) pathway, an important mechanism involved in the malignant transformation of several different tumors, also plays an eminent role in EC progression [102]. In this respect, in addition to the association between greater EGF and MMP-9 expression and a more invasive phenotype observed in EC tumors [103], it is known that ESCC cell line treatment with recombinant EGF leads to MMP-9 expression enhancement [104]. Of note, the study of Okawa and colleagues reported that the crosstalk between epidermal growth factor receptor (EGFR), human telomerase reverse transcriptase (hTERT), and p53 are directly associated with invasion of stromal compartment through the activation of MMP-9, but not that of MMP-2 [105]. Therefore, instead of MAPK signaling pathway, which seems to represent a central pathway involved in the regulation of MMP-2 and MMP-9, EGF signaling pathway likely participates strictly in the regulation of MMP-9, with these mechanisms being associated with PI3K activation and p53 cooperation [105,106].

2.3. Glycoproteins and ECM Adhesion and Migration

The activation of key cellular events depends on the interaction between cells and ECM adhesion molecules, which consists of a central mechanism represented not only by the "adhesion process itself", but also by the activation of several signaling cascades that trigger crucial behaviors involved in the maintenance of tissue homeostasis and cancer development [107,108]. In this way, loss of E-cadherin, which plays a central role in cellular adhesion and communication by primarily mediating cell–cell adhesion, during tumor progression is directly associated with invasiveness and metastatic potential [109]. Moreover, classic malignant behaviors associated with EC progression, such as EMT, are also linked to decreased or lacking functional E-cadherin [110]. Particularly in EC, E-cadherin has drawn attention due to its great potential role as a prognostic biomarker. A meta-analysis study suggested that decreased levels of E-cadherin-positive staining are typical of undifferentiated tumor cells, and it has been proposed as a prognostic marker for ESCC patients [111]. Additionally, it was demonstrated that downregulation of E-cadherin by EC cells was directly correlated with increased risk of lymph node metastasis and advanced clinical stage [111]. Although molecular mechanisms are unclear, some target genes have been investigated and linked to reduction of E-cadherin and ESCC progression, such as p21 and cyclooxygenase-2 (COX-2) [112]. In this regard, these observations reinforce the notion that adhesion proteins, such as integrins, cadherins, and its ECM partners, play an essential role in esophageal carcinogenesis. In fact, the interaction between tumor cells and ECM mediated by adhesion glycoproteins could force the anomalous progression of cell cycle by tumor cells due to the activation of a complex circuit, which culminates in the decreased expression of the cell cycle regulators p15 and p21 [113]. Moreover, these data corroborate the previously reported association between aberrant expression of adhesion proteins, such as laminin and fibronectin, and EC progression [114,115]. Laminins are high molecular weight heterotrimeric glycoproteins produced by the association between α, β, and γ chains, playing a fundamental role for the physiology of basement membranes [116]. In fact, it has been reported that esophageal basaloid-squamous cell carcinomas displayed positive laminin and type IV collagen staining, being basement membranes thinner than in healthy esophageal tissues, and occasionally discontinuous in the EC [117]. Additionally, it has been reported that laminin receptors are progressively overexpressed in ESCC from stage I to III, such as a 67 kDa laminin receptor [114], which reinforces the association of this glycoprotein and ESCC prognosis. Nonetheless, another study demonstrated that in invasive tumors, basement membranes were partially or completely lost, depending on inflammatory pattern and epithelial organization [118]. The correlation between lack of basement membranes, inflammatory reaction in situ, and EC progression was reinforced by studies that linked these findings with secretion of proteolytic enzymes produced by cancer cells [119]. Since then, different subunits of laminins have been described and associated with the prognosis of patients affected by EC. For instance, γ2 chain of laminin-5 (Lam5 (γ2)), composed by the α3, β3, and γ2 chains, is frequently detected at elevated levels in invasive carcinomas and widely linked to recurrence in ESCC patients [120]. Indeed, co-expression of Lam5 (γ2) chain and EGFR indicates a very poor prognosis of ESCC due to high metastatic potential [121]. In fact, it seems that a mutual relationship between Lam5 (γ2) and EGF pathway exists along EC development and, besides the fact that EGF enhances Lam5 (γ2) expression [122], degradation of γ2 subunit by MT1-MMP produces a functional biological fragment, called DIII, which is able to bind and activate the EGFR, culminating with an increase in MMP-2 expression and cellular migration [123]. In accordance with these data, when in contact with MMP-1, Lam5 (γ2) cleavage enhances invasiveness and metastasis in ESCC patients [124], thus corroborating the biological consequences associated with Lam5 (γ2) degradation during ESCC progression. Moreover, laminin-5 β3 chain expression is significantly elevated in ESCC tissues in comparison with healthy esophageal tissues. This finding is directly correlated with enhanced metastatic potential and decreased life expectance in EC patients [125]. Concerning laminin-5, this protein is located in the basement membrane compartment of EC and seems to play a relevant role in ESCC carcinogenesis because it enhances tumor invasiveness potential by activating the PI3K pathway [126]. It has been demonstrated that ECM plays regulatory roles in distinct events during tumorigenesis and cancer progression; therefore, unveiling ECM composition is critical to deeper understand the molecular mechanisms through which it controls such crucial events and, consequently, to envisage potential new therapeutic strategies. Recently, Senthebane and colleagues investigated the chemotherapeutic response of ESCC cell lines cultured over 3D cell-derived ECM. In this sense, it was observed that differential expression and deposition of ECM proteins, such as collagens, fibronectin, and laminins, by ESCC cell lines was able to decrease the efficacy of different chemotherapeutic agents [38]. Further, high concentration of ECM components in the serum, including hyaluronic acid (HA) and laminin, was suggested as a potential marker for assessing upper gastrointestinal cancer development risk [127]. In EAC, high expression of laminin is associated with enhanced cell detachment, invasion, and dissemination in an osteopontin (OPN)-dependent manner [128]. Finally, the expression of laminins in EC can also be regulated by miRNAs. In this sense, it has been recently reported that laminin α1, which is widely distributed in ESCC tissues, displays an inverse correlation with the expression of miRNA-202. In other words, miRNA-202 is capable of suppressing tumor progression by targeting laminin α1 in ESCC, and high levels of laminin α1 have been associated with poor prognosis [129]. Moreover, these data are in accordance with miRNA-202 low expression observed in ESCC tissue [130,131], which was significantly associated with local invasion and metastasis [131]. Focusing on another important ECM adhesion protein, it has been already described that fibronectin is highly associated with esophageal tumorigenesis and esophageal patients' prognosis; nevertheless, the mechanisms through which it occurs are still unclear [132]. Fibronectin is a glycoprotein present in the ECM of normal and cancerous tissues, including the esophagus, where it regulates cellular events such as adhesion, differentiation, proliferation, and metastasis. In ESCC, fibronectin is predominantly found in tumor stroma and is widely associated with lymphatic metastasis, indicating that fibronectin exerts stromal regulatory functions favoring tumor cell metastasis [133]. Inflammatory conditions also interfere with ECM organization in EC. In ESCC cells, the exposure to lipopolysaccharide (LPS) potentiated cell adhesion to fibronectin and hepatic sinusoids through toll-like receptor 4 (TLR4) signaling and selectin ligands [134]. Further, EAC-associated risk factors are capable of inducing TLR4 expression in a normal esophageal epithelium cell line, as well as in EAC cell lines (data not published). In fact, these observations sound very interesting, and the association between inflammatory process and EC development has emerged as a key event along disease development [39,135]. Furthermore, in silico analysis identified genes and signaling pathways potentially involved with EAC genesis and development, with it being detected that fibronectin 1 may represent a major target associated with metastatic capacity [136]. Indeed, fibronectin is upregulated in EAC [137], but nevertheless its contribution to cell invasion and migration is still under investigation. In this context, several authors have contributed with mechanistic proposals. One of them includes the transcriptional network of sex-determining region Y-box transcription factor 17 (SOX17) regulating fibronectin functions in ESCC tissues. Approximately 47% of ESCC patients present low levels of SOX17, which inhibits the expression of fibronectin and other genes involved with ESCC progression and metastasis [138]. On the other hand, it was demonstrated that the overexpression of fibronectin in ESCC cells could be associated with the activation of the MAPK pathway [115], as well as with miRNA regulation [139]. In this way, in ESCC patients, low miRNA-1 expression hallmarks poor clinical outcome and lymph node metastasis, whereas high expression of miRNA-1 promotes apoptosis followed by cell cycle arrest in ESCC cells, as well as fibronectin 1 organization [139]. Additionally, fibronectin may also be regulated by long noncoding RNAs (lncRNAs). It has been already reported that co-overexpression of fibronectin 1 and lncRNAs, such as lnc-ABCA12-3 (ATP Binding Cassette Subfamily A Member 12-3), in ESCC tissues is associated with tumor expansion, metastasis, and patients' poor prognosis [140]. Furthermore, increased expression of lncRNA SPRY4-IT1 (sprouty RTK signaling antagonist 4—Intronic Transcript 1) observed in ESCC cell lines increased cell motility and favored EMT, an event strongly related to an increase in the cell motility that could be corroborated by fibronectin-positive staining [141]. In addition, during EMT-induced by hypoxia, fibronectin upregulation was directly associated with hypoxia-inducible factor 1-α (HIF-1α) secretion and downregulation of E-cadherin [142]. These data seem interesting, as hypoxia is a phenomenon largely associated with the tumor microenvironment, reinforcing the functional association of fibronectin and EMT phenotype as a tumor progression mechanism triggered by hypoxic context. Another ECM member glycoprotein involved in EC genesis and/or progression is tenascin-C. This protein controls proliferation, differentiation, and migration of different cell types in normal and tumoral tissues. In ESCC patients, this glycoprotein is predominantly detected surrounding tumor stromal cells, invading the submucosa [143]. Tenascin-C is considered as a predictive marker for EAC progression [144], and it is associated with ESCC patients' poor prognosis due to its capacity of favoring invasion and metastasis [145,146]. In EAC, upregulation and co-overexpression of tenascin-C and fibronectin have been directly associated with poor prognosis and metastasis of the patients [137]. The mechanisms that regulate tenascin-C expression and its involvement with the development of EC remain unclear, although some efforts have already been made to shed some light on this question. Recently, Yang and colleagues demonstrated that tenascin-C displays potential in amplifying cancer stem-like properties, at least in part, indicating that tenascin-C may be involved with tumor recurrence [147]. These authors revealed that expression of tenascin-C in ESCC cells is directly associated with that of SRY-box transcription factor 2 (SOX2), a well-known cancer stem cell marker. Further, they also demonstrated that tenascin-C promotes EMT through an Akt/ Hypoxia Inducible Factor 1 Subunit Alpha (HIF1-α)-dependent manner [147]. Finally, in order to investigate ECM components with potential to regulate cellular and molecular mechanisms involved with carcinogenesis, attention has been drawn to glycoproteins that recognize membrane compartments on tumoral cells and structural ECM proteins. One of them is galectin-3, a β-galactoside-binding protein that regulates cell–cell and cell–ECM interactions and is highly associated with tumorigenesis and cancer progression [148]. In the tumor microenvironment, extracellular galectin-3 interacts with distinct ECM molecules and surface glycoproteins, such as growth factor receptors, integrins, cadherins, and members of Notch family, probably favoring the biology of cancer cells [149]. Galectin-3 interacts with galectin-3-binding protein (LGALS3BP), a receptor/ligand of galectin-3, other galectins, and ECM compounds, such as collagens and fibronectin. In ESCC cells, LGALS3BP is overexpressed in most tumors [150]. On the other hand, LGALS3BP is not considered a specific marker for ESCC progression [151]. In EC cell line Eca109, downregulation of galectin-3 induced pro-apoptotic mechanisms and inhibited proliferation, migration, and invasion [152]. Additionally, Eca109 and EC9706 cells reduced the capacity of forming tubular networks upon galectin-3 knockdown in a phenomenon associated with MMP-2 downmodulation upon galectin-3 silencing [153]. Furthermore, ESCC cell lines lacking galectin-3 expression display lower viability, mitotic index, and invasiveness capacity compared with control cells, at least in part, as a consequence of reduced EGFR endocytosis [154]. Other galectins have been linked to EC formation and progression. For instance, galectin-9 enhances antitumor immunity mediated by Tim-3[+] dendritic cells and CD8[+] T cells [155]. In EC cell lines, galectin-9 inhibited cell proliferation in a concentration-dependent manner and induced mitochondria-mediated apoptosis [156]. Moreover, galectin-9 showed antitumor effects in vitro, inducing apoptosis followed by high levels of caspase-cleaved cytokeratin 18, activated caspase-3, and activated caspase-9 [157]. In addition, galectin-7 overexpression in ESCC tissues has been suggested as a potential biomarker for ESCC diagnosis [158]. Furthermore, it has also been reported that some proteoglycans, such as dermatan sulfate and hyaluronan, are overexpressed in ESCC tissue [159,160]. Finally, its deregulated expression is capable of altering not only the adhesion properties of EC cells, but also of improving EC cell migration and invasion abilities through the activation of phosphorylated extracellular signal-regulated kinases (pErk1/2) and FAK [160,161].

3. Conclusions

ECM plays regulatory roles in the tumor microenvironment controlling tumorigenesis, progression, and metastasis. There are several biological events intrinsically linked to ECM that are crucial for cancer formation and progression, including cell differentiation and proliferation, apoptosis evasion, immunological disturbances, and inflammation. The reciprocal interactions between ECM and the surrounding EC cells may represent a fundamental step along disease development and progression, as described in Figure 2. However, the vast repertory of interactions between EC cells and ECM seems to be complex, and the mechanisms involved during the establishment of each interaction could be regulated by dozens or even hundreds of different proteins. Moreover, the singular natural history of EC, which is strongly characterized by the influence of distinct ethiopathological factors, improves the complexity of the current scenario. However, due to the scarce knowledge on the cellular and molecular mechanisms involved in this disease development, the growing evidence of ECM's role along esophageal carcinogenesis could provide a solid base to improve EC management in the future.

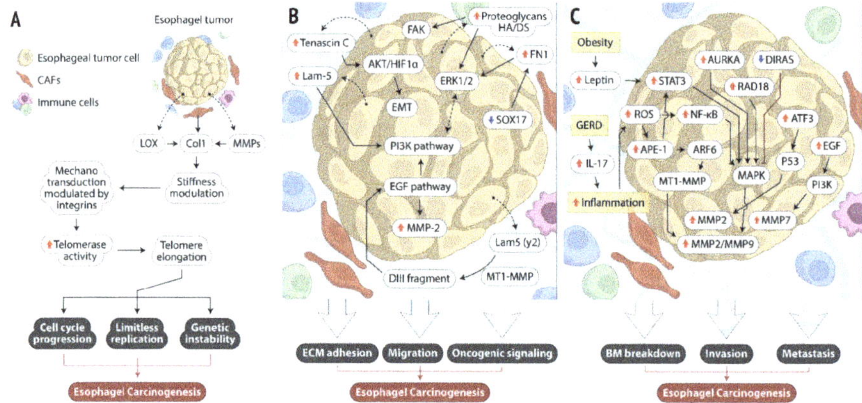

Figure 2. Mechanistic pathways involved in the reciprocal interactions between esophageal cancer (EC) cells and extracellular matrix (ECM). The figure illustrates the main molecules and signaling pathways altered along the interaction between tumor cells and ECM, which may represent fundamental steps for esophageal carcinogenesis. (**A**) Lysyl oxidase (LOX)- and matrix metalloproteinase (MMP)-altered expression in esophageal tumors is involved in the turnover of collagen molecules, resulting in ECM stiffness modulation that, in turn, may enhance telomerase activity, resulting in activation of signaling pathways involved with malignant phenotype acquisition, such as cell cycle progression, limitless replication, and genetic instability. (**B**) Esophageal tumor cell release increased levels of ECM adhesion molecules that activate intracellular signaling pathways involved with cell survival, enhanced cellular migration, and activation of oncogenic signaling. (**C**) Risk factors associated with esophageal tumor development (obesity and gastro-esophageal reflux disease—GERD) lead to an increase in leptin and interleukin 17 (IL-17) levels, which generates an inflammatory environment. The alterations in leptin and inflammation levels modulate the expression of several intracellular molecules that, together with other altered signaling pathways, converge to the activation of MMP expression and activity, entailing basement membrane (BM) breakdown, cellular invasion, and metastasis. Dotted arrows: molecules secreted by esophageal tumor cells; solid arrows: activation of cellular events or intracellular signaling pathways; red solid arrow: overexpressed molecules; solid blue arrow: underexpressed molecule.

Author Contributions: A.P.J. and L.E.N. conceived, delineated, discussed, and wrote the manuscript. N.M.D.C. and L.F.R.P. discussed, critically reviewed, and wrote the manuscript. B.P., F.L.d.O., and M.L.C. contributed to the discussion and wrote the manuscript. All authors have read and agreed to the published version of the manuscript.

Funding: This research was funded by Conselho Nacional de Desenvolvimento Científico e Tecnológico (CNPq—Brazil), Coordenação de Aperfeiçoamento de Pessoal de Nível Superior (CAPES), Fundação de Amparo à Pesquisa Carlos Chagas Filho (FAPERJ—Brazil), Programa de Oncobiologia IBqM, Fundação Para Pesquisa e Estudo do Câncer Ary Frauzino (FAF), and Swiss Bridge Foundation. The APC was funded by grants received by A.P.J. and L.E.N.

Conflicts of Interest: The authors declare that they have no competing interests.

Abbreviations

α-SMA	α-smooth muscle actin
ABCA12	3 ATP Binding Cassette Subfamily A Member 12-3
Akt	protein kinase B
AP-1	activator protein 1
APE1	apurinic/apyrimidinic endonuclease 1
ARF6	ADP-ribosylation factor 6

ATF3	activating transcription factor 3
AURKA	Aurora A kinase
BE	Barrett's esophagus
CAF	cancer-associated fibroblasts
COX-2	cyclooxygenase-2
CXCR4	C-X-C motif chemokine receptor 4
DIRAS1	Distinct Subgroup of The Ras Family Member 1
DDR	discoidin domain receptor
EAC	esophageal adenocarcinoma
EC	esophageal cancer
ECM	extracellular matrix
EGF	epidermal growth factor
EGFR	epidermal growth factor receptor
EMT	epithelial mesenchymal transition
Erk	extracellular-signal regulated kinase
ESCC	esophageal squamous cell carcinoma
FAK	focal adhesion kinase
FAP-α	fibroblast activation protein α
GERD	gastroesophageal reflux disease
HA	hyaluronic acid
HIF-1α	hypoxia-inducible factor 1-α
hTERT	human telomerase reverse transcriptase
IL-17	interleukin 17
JNK	c-Jun N-terminal kinase
Lam5 (γ2)	chain γ2 of laminin-5
LGALS3BP	galectin-3 binding protein
lncRNA	long noncoding RNA
LOX	lysyl oxidase
LPS	lipopolysaccharide
MAPK	mitogen-activated protein kinase
miRNA	microRNA
MMP	matrix metalloproteinase
MT1-MMP	membrane-type 1 matrix metalloproteinase
NF-kB	nuclear factor kappa B
OPN	osteopontin
P38	p38 mitogen-activated protein kinases
pErk	phosphorylated extracellular signal-regulated kinases
PI3K	phosphoinositide 3-kinase
PTK7	protein tyrosine kinase 7
ROS	reactive oxygen species
SOX	sex-determining region Y-box-box transcription factor
SPRY4-IT1	sprouty RTK signaling antagonist 4—Intronic Transcript 1
STAT3	signal transducer and activator of transcription 3
TAF	tumor associated fibroblasts
TLR4	toll-like receptor 4
VEGF	vascular endothelial growth factor

References

1. Frantz, C.; Stewart, K.M.; Weaver, V.M. The extracellular matrix at a glance. *J. Cell Sci.* **2010**, *123*, 4195–4200. [CrossRef] [PubMed]
2. Huang-Doran, I.; Zhang, C.-Y.; Vidal-Puig, A. Extracellular Vesicles: Novel Mediators of Cell Communication in Metabolic Disease. *Trends Endocrinol. Metab.* **2017**, *28*, 3–18. [CrossRef] [PubMed]
3. Gjorevski, N.; Nelson, C.M. Bidirectional extracellular matrix signaling during tissue morphogenesis. *Cytokine Growth Factor Rev.* **2009**, *20*, 459–465. [CrossRef] [PubMed]

4. Plotnikov, E.Y.; Silachev, D.N.; Popkov, V.A.; Zorova, L.D.; Pevzner, I.B.; Zorov, S.D.; Jankauskas, S.S.; Babenko, V.A.; Sukhikh, G.T.; Zorov, D.B. Intercellular Signalling Cross-Talk: To Kill, To Heal and To Rejuvenate. *Heart Lung Circ.* **2017**, *26*, 648–659. [CrossRef]

5. Bosman, F.T.; Stamenkovic, I. Functional structure and composition of the extracellular matrix. *J. Pathol.* **2003**, *200*, 423–428. [CrossRef]

6. Hynes, R.O. Extracellular matrix: Not just pretty fibrils. *Science* **2009**, *326*, 1216–1219. [CrossRef]

7. Nelson, C.M.; Bissell, M.J. Of extracellular matrix, scaffolds, and signaling: Tissue architecture regulates development, homeostasis, and cancer. *Annu. Rev. Cell Dev. Biol.* **2006**, *22*, 287–309. [CrossRef]

8. Pickup, M.W.; Mouw, J.K.; Weaver, V.M. The extracellular matrix modulates the hallmarks of cancer. *EMBO Rep.* **2014**, *15*, 1243–1253. [CrossRef]

9. Ghajar, C.M.; Bissell, M.J. Extracellular matrix control of mammary gland morphogenesis and tumorigenesis: Insights from imaging. *Histochem. Cell Biol.* **2008**, *130*, 1105–1118. [CrossRef]

10. Pearce, O.M.; Delaine-Smith, R.M.; Maniati, E.; Nichols, S.; Wang, J.; Böhm, S.; Rajeeve, V.; Ullah, D.; Chakravarty, P.; Jones, R.R.; et al. Deconstruction of a Metastatic Tumor Microenvironment Reveals a Common Matrix Response in Human Cancers. *Cancer Discov.* **2017**, *8*, 304–319. [CrossRef]

11. Yuzhalin, A.E.; Urbonas, T.; Silva, M.A.; Muschel, R.J.; Gordon-Weeks, A.N. A core matrisome gene signature predicts cancer outcome. *Br. J. Cancer* **2018**, *118*, 435–440. [CrossRef] [PubMed]

12. Kai, F.; Drain, A.P.; Weaver, V.M. The Extracellular Matrix Modulates the Metastatic Journey. *Dev. Cell* **2019**, *49*, 332–346. [CrossRef]

13. Comoglio, P.M.; Trusolino, L. Cancer: The matrix is now in control. *Nat. Med.* **2005**, *11*, 1156–1158. [CrossRef] [PubMed]

14. Lu, P.; Weaver, V.M.; Werb, Z. The extracellular matrix: A dynamic niche in cancer progression. *J. Cell Biol.* **2012**, *196*, 395–406. [CrossRef] [PubMed]

15. Lin, E.W.; Karakasheva, T.A.; Hicks, P.D.; Bass, A.J.; Rustgi, A.K. The tumor microenvironment in esophageal cancer. *Oncogene* **2016**, *35*, 5337–5349. [CrossRef]

16. Bray, F.; Ferlay, J.; Soerjomataram, I.; Siegel, R.L.; Torre, L.A.; Jemal, A. Global cancer statistics 2018: GLOBOCAN estimates of incidence and mortality worldwide for 36 cancers in 185 countries. *CA Cancer J. Clin.* **2018**, *68*, 394–424. [CrossRef]

17. Pennathur, A.; Gibson, M.K.; Jobe, B.A.; Luketich, J.D. Oesophageal carcinoma. *Lancet* **2013**, *381*, 400–412. [CrossRef]

18. Lagergren, J.; Lagergren, P. Oesophageal cancer. *BMJ* **2010**, *341*, c6280. [CrossRef]

19. Coleman, H.G.; Xie, S.-H.; Lagergren, J. The Epidemiology of Esophageal Adenocarcinoma. *Gastroenterology* **2018**, *154*, 390–405. [CrossRef]

20. Zhang, Y. Epidemiology of esophageal cancer. *World J. Gastroenterol.* **2013**, *19*, 5598–5606. [CrossRef]

21. Gupta, B.; Kumar, N. Worldwide incidence, mortality and time trends for cancer of the oesophagus. *Eur. J. Cancer Prev.* **2017**, *26*, 107–118. [CrossRef]

22. Kalluri, R.; Zeisberg, M. Fibroblasts in cancer. *Nat. Rev. Cancer* **2006**, *6*, 392–401. [CrossRef] [PubMed]

23. Saito, S.; Morishima, K.; Ui, T.; Matsubara, D.; Tamura, T.; Oguni, S.; Hosoya, Y.; Sata, N.; Lefor, A.T.; Yasuda, Y.; et al. Stromal fibroblasts are predictors of disease-related mortality in esophageal squamous cell carcinoma. *Oncol. Rep.* **2014**, *32*, 348–354. [CrossRef] [PubMed]

24. Kalluri, R. The biology and function of fibroblasts in cancer. *Nat. Rev. Cancer* **2016**, *16*, 582–598. [CrossRef] [PubMed]

25. Kretschmer, I.; Freudenberger, T.; Twarock, S.; Yamaguchi, Y.; Grandoch, M.; Fischer, J.W. Esophageal squamous cell carcinoma cells modulate chemokine expression and hyaluronan synthesis in fibroblasts. *J. Biol. Chem.* **2016**, *291*, 4091–4106. [CrossRef] [PubMed]

26. Wang, J.; Zhang, G.; Wang, J.; Wang, L.; Huang, X.; Cheng, Y. The role of cancer-associated fibroblasts in esophageal cancer. *J. Transl. Med.* **2016**, *14*, 30. [CrossRef] [PubMed]

27. Fan, D.; Creemers, E.E.; Kassiri, Z. Matrix as an Interstitial Transport System. *Circ. Res.* **2014**, *114*, 889–902. [CrossRef]

28. Marks, D.L.; Olson, R.L.; Fernandez-Zapico, E.M. Epigenetic control of the tumor microenvironment. *Epigenomics* **2016**, *8*, 1671–1687. [CrossRef]

29. Fang, M.; Yuan, J.; Peng, C.; Li, Y. Collagen as a double-edged sword in tumor progression. *Tumour Biol.* **2014**, *35*, 2871–2882. [CrossRef]

30. Northcott, J.M.; Dean, I.S.; Mouw, J.K.; Weaver, V.M. Feeling Stress: The Mechanics of Cancer Progression and Aggression. *Front. Cell Dev. Biol.* **2018**, *6*, 17. [CrossRef]

31. Mouw, J.K.; Ou, G.; Weaver, V.M. Extracellular matrix assembly: A multiscale deconstruction. *Nat. Rev. Mol. Cell Biol.* **2014**, *15*, 771–785. [CrossRef] [PubMed]

32. Paszek, M.J.; Zahir, N.; Johnson, K.R.; Lakins, J.N.; Rozenberg, G.I.; Gefen, A.; Reinhart-King, C.A.; Margulies, S.S.; Dembo, M.; Boettiger, D.; et al. Tensional homeostasis and the malignant phenotype. *Cancer Cell* **2005**, *8*, 241–254. [CrossRef] [PubMed]

33. Rudnick, J.A.; Kuperwasser, C. Stromal biomarkers in breast cancer development and progression. *Clin. Exp. Metastasis* **2012**, *29*, 663–672. [CrossRef]

34. Voiles, L.; Lewis, D.E.; Han, L.; Lupov, I.P.; Lin, T.-L.; Robertson, M.J.; Petrache, I.; Chang, H.-C. Overexpression of type VI collagen in neoplastic lung tissues. *Oncol. Rep.* **2014**, *32*, 1897–1904. [CrossRef] [PubMed]

35. Xiong, G.; Deng, L.; Zhu, J.; Rychahou, P.G.; Xu, R. Prolyl-4-hydroxylase α subunit 2 promotes breast cancer progression and metastasis by regulating collagen deposition. *BMC Cancer* **2014**, *14*, 1. [CrossRef] [PubMed]

36. Zhu, J.; Xiong, G.; Trinkle, C.; Xu, R. Integrated extracellular matrix signaling in mammary gland development and breast cancer progression. *Histol. Histopathol.* **2014**, *29*, 1083–1092.

37. Troester, M.A.; Lee, M.H.; Carter, M.; Fan, C.; Cowan, D.W.; Perez, E.R.; Pirone, J.R.; Perou, C.M.; Jerry, D.J.; Schneider, S.S. Activation of Host Wound Responses in Breast Cancer Microenvironment. *Clin. Cancer Res.* **2009**, *15*, 7020–7028. [CrossRef]

38. Senthebane, D.A.; Jonker, T.; Rowe, A.; Thomford, N.E.; Munro, D.; Dandara, C.; Wonkam, A.; Govender, D.; Calder, B.; Soares, N.C.; et al. The Role of Tumor Microenvironment in Chemoresistance: 3D Extracellular Matrices as Accomplices. *Int. J. Mol. Sci.* **2018**, *19*, 2861. [CrossRef]

39. Hanahan, D.; Weinberg, R.A. Hallmarks of Cancer: The Next Generation. *Cell* **2011**, *144*, 646–674. [CrossRef]

40. Wintzell, M.; Hjerpe, E.; Lundqvist, E.Å.; Shoshan, M. Protein markers of cancer-associated fibroblasts and tumor-initiating cells reveal subpopulations in freshly isolated ovarian cancer ascites. *BMC Cancer* **2012**, *12*, 359. [CrossRef]

41. Öhlund, D.; Handly-Santana, A.; Biffi, G.; Elyada, E.; Almeida, A.S.; Ponz-Sarvise, M.; Corbo, V.; Oni, T.E.; Hearn, S.A.; Lee, E.J.; et al. Distinct populations of inflammatory fibroblasts and myofibroblasts in pancreatic cancer. *J. Exp. Med.* **2017**, *214*, 579–596. [CrossRef] [PubMed]

42. Mitra, A.K.; Zillhardt, M.; Hua, Y.; Tiwari, P.; Murmann, A.E.; Peter, M.E.; Lengyel, E. MicroRNAs reprogram normal fibroblasts into cancer-associated fibroblasts in ovarian cancer. *Cancer Discov.* **2012**, *2*, 1100–1108. [CrossRef]

43. Tanaka, K.; Miyata, H.; Sugimura, K.; Fukuda, S.; Kanemura, T.; Yamashita, K.; Miyazaki, Y.; Takahashi, T.; Kurokawa, Y.; Yamasaki, M.; et al. miR-27 is associated with chemoresistance in esophageal cancer through transformation of normal fibroblasts to cancer-associated fibroblasts. *Carcinogenesis* **2015**, *36*, 894–903. [CrossRef] [PubMed]

44. Orimo, A.; Gupta, P.B.; Sgroi, D.C.; Arenzana-Seisdedos, F.; Delaunay, T.; Naeem, R.; Carey, V.J.; Richardson, A.L.; Weinberg, R.A. Stromal Fibroblasts Present in Invasive Human Breast Carcinomas Promote Tumor Growth and Angiogenesis through Elevated SDF-1/CXCL12 Secretion. *Cell* **2005**, *121*, 335–348. [CrossRef]

45. Cat, B.; Stuhlmann, D.; Steinbrenner, H.; Alili, L.; Holtkötter, O.; Sies, H.; Brenneisen, P. Enhancement of tumor invasion depends on transdifferentiation of skin fibroblasts mediated by reactive oxygen species. *J. Cell Sci.* **2006**, *119*, 2727–2738. [CrossRef]

46. Marsh, D.; Suchak, K.; Moutasim, K.A.; Vallath, S.; Hopper, C.; Jerjes, W.; Upile, T.; Kalavrezos, N.; Violette, S.M.; Weinreb, P.H.; et al. Stromal features are predictive of disease mortality in oral cancer patients. *J. Pathol.* **2011**, *223*, 470–481. [CrossRef]

47. Quail, D.F.; Joyce, J.A. Microenvironmental regulation of tumor progression and metastasis. *Nat. Med.* **2013**, *19*, 1423–1437. [CrossRef]

48. Torres, S.; Bartolomé, R.A.; Mendes, M.; Barderas, R.; Fernandez-Aceñero, M.J.; Peláez-García, A.; Peña, C.; Lopez-Lucendo, M.; Villar-Vázquez, R.; De Herreros, A.G.; et al. Proteome Profiling of Cancer-Associated Fibroblasts Identifies Novel Proinflammatory Signatures and Prognostic Markers for Colorectal Cancer. *Clin. Cancer Res.* **2013**, *19*, 6006–6019. [CrossRef]

49. Ishibashi, N.; Maebayashi, T.; Aizawa, T.; Sakaguchi, M.; Okada, M. Serum tumor marker levels at the development of intracranial metastasis in patients with lung or breast cancer. *J. Thorac. Dis.* **2019**, *11*, 1765–1771. [CrossRef]

50. Palmieri, D.; Astigiano, S.; Barbieri, O.; Ferrari, N.; Marchisio, S.; Ulivi, V.; Volta, C.; Manduca, P. Procollagen I COOH-terminal fragment induces VEGF-A and CXCR4 expression in breast carcinoma cells. *Exp. Cell Res.* **2008**, *314*, 2289–2298. [CrossRef]

51. Jabłońska-Trypuć, A.; Matejczyk, M.; Rosochacki, S. Matrix metalloproteinases (MMPs), the main extracellular matrix (ECM) enzymes in collagen degradation, as a target for anticancer drugs. *J. Enzym. Inhib. Med. Chem.* **2016**, *31*, 177–183. [CrossRef]

52. Fang, S.; Dai, Y.; Mei, Y.; Yang, M.; Hu, L.; Yang, H.; Guan, X.; Li, J. Clinical significance and biological role of cancer-derived Type I collagen in lung and esophageal cancers. *Thorac. Cancer* **2019**, *10*, 277–288. [CrossRef] [PubMed]

53. Li, J.; Wang, X.; Zheng, K.; Liu, Y.; Li, J.; Wang, S.; Liu, K.; Song, X.; Li, N.; Xie, S.; et al. The clinical significance of collagen family gene expression in esophageal squamous cell carcinoma. *PeerJ* **2019**, *7*, e7705. [CrossRef] [PubMed]

54. Hanley, C.J.; Noble, F.; Ward, M.; Bullock, M.; Drifka, C.; Mellone, M.; Manousopoulou, A.; Johnston, H.E.; Hayden, A.; Thirdborough, S.; et al. A subset of myofibroblastic cancer-associated fibroblasts regulate collagen fiber elongation, which is prognostic in multiple cancers. *Oncotarget* **2016**, *7*, 6159–6174. [CrossRef] [PubMed]

55. Rammal, H.; Saby, C.; Magnien, K.; Van-Gulick, L.; Garnotel, R.; Buache, E.; El Btaouri, H.; Jeannesson, P.; Morjani, H. Discoidin Domain Receptors: Potential actors and targets in cancer. *Front. Pharmacol.* **2016**, *7*, 55. [CrossRef] [PubMed]

56. Croissant, C.; Tuariihionoa, A.; Bacou, M.; Souleyreau, W.; Sala, M.; Henriet, E.; Bikfalvi, A.; Saltel, F.; Auguste, P.; Bikvalvi, A. DDR1 and DDR2 physical interaction leads to signaling interconnection but with possible distinct functions. *Cell Adhes. Migr.* **2018**, *12*, 324–334. [CrossRef]

57. Huang, Y.; Arora, P.; McCulloch, C.A.; Vogel, W.F. The collagen receptor DDR1 regulates cell spreading and motility by associating with myosin IIA. *J. Cell Sci.* **2009**, *122*, 1637–1646. [CrossRef]

58. Ruiz, P.A.; Jarai, G. Collagen I Induces Discoidin Domain Receptor (DDR) 1 Expression through DDR2 and a JAK2-ERK1/2-mediated Mechanism in Primary Human Lung Fibroblasts. *J. Biol. Chem.* **2011**, *286*, 12912–12923. [CrossRef]

59. Fong, S.F.T.; Dietzsch, E.; Fong, K.S.K.; Hollósi, P.; Asuncion, L.; He, Q.; Parker, M.I.; Csiszar, K. Lysyl oxidase-like 2 expression is increased in colon and esophageal tumors and associated with less differentiated colon tumors. *Genes Chromosom. Cancer* **2007**, *46*, 644–655. [CrossRef]

60. Pylayeva, Y.; Gillen, K.M.; Gerald, W.; Beggs, H.E.; Reichardt, L.F.; Giancotti, F.G. Ras- and PI3K-dependent breast tumorigenesis in mice and humans requires focal adhesion kinase signaling. *J. Clin. Investig.* **2009**, *119*, 252–266. [CrossRef]

61. Matte, B.F.; Kumar, A.; Placone, J.K.; Zanella, V.G.; Martins, M.D.; Engler, A.J.; Lamers, M.L. Matrix stiffness mechanically conditions EMT and migratory behavior of oral squamous cell carcinoma. *J. Cell Sci.* **2019**, *132*, jcs224360. [CrossRef] [PubMed]

62. Yu, S.T.; Chen, L.; Wang, H.J.; Tang, X.D.; Fang, D.C.; Yang, S.M. hTERT promotes the invasion of telomerase-negative tumor cells in vitro. *Int. J. Oncol.* **2009**, *35*, 329–336. [PubMed]

63. Pal, J.; Gold, J.S.; Munshi, N.C.; Shammas, M.A. Biology of telomeres: Importance in etiology of esophageal cancer and as therapeutic target. *Transl. Res.* **2013**, *162*, 364–370. [CrossRef] [PubMed]

64. Kunimura, C.; Kikuchi, K.; Ahmed, N.; Shimizu, A.; Yasumoto, S. Telomerase activity in a specific cell subset co-expressing integrin beta1/EGFR but not p75NGFR/bcl2/integrin beta4 in normal human epithelial cells. *Oncogene* **1998**, *17*, 187–197. [CrossRef]

65. Vay, C.; Hosch, S.B.; Stoecklein, N.H.; Klein, C.A.; Vallbohmer, D.; Link, B.-C.; Yekebas, E.F.; Izbicki, J.R.; Knoefel, W.T.; Scheunemann, P. Integrin Expression in Esophageal Squamous Cell Carcinoma: Loss of the Physiological Integrin Expression Pattern Correlates with Disease Progression. *PLoS ONE* **2014**, *9*, e109026. [CrossRef]

66. Kai, F.; Laklai, H.; Weaver, V.M. Force Matters: Biomechanical Regulation of Cell Invasion and Migration in Disease. *Trends Cell Biol.* **2016**, *26*, 486–497. [CrossRef]

67. Wolf, K.; Lindert, M.T.; Krause, M.; Alexander, S.; Riet, J.T.; Willis, A.L.; Hoffman, R.M.; Figdor, C.G.; Weiss, S.J.; Friedl, P. Physical limits of cell migration: Control by ECM space and nuclear deformation and tuning by proteolysis and traction force. *J. Cell Biol.* **2013**, *201*, 1069–1084. [CrossRef]

68. Delcommenne, M.; Streuli, C.H. Control of Integrin Expression by Extracellular Matrix. *J. Biol. Chem.* **1995**, *270*, 26794–26801. [CrossRef]

69. Nukuda, A.; Sasaki, C.; Ishihara, S.; Mizutani, T.; Nakamura, K.; Ayabe, T.; Kawabata, K.; Haga, H. Stiff substrates increase YAP-signaling-mediated matrix metalloproteinase-7 expression. *Oncogenesis* **2015**, *4*, e165. [CrossRef]

70. McGrail, D.J.; Kieu, Q.M.N.; Iandoli, J.A.; Dawson, M.R. Actomyosin tension as a determinant of metastatic cancer mechanical tropism. *Phys. Biol.* **2015**, *12*, 026001. [CrossRef]

71. Bonnans, C.; Chou, J.; Werb, Z. Remodelling the extracellular matrix in development and disease. *Nat. Rev. Mol. Cell Biol.* **2014**, *15*, 786–801. [CrossRef]

72. Lei, Z.; Jian, M.; Li, X.; Wei, J.; Meng, X.; Wang, Z. Biosensors and bioassays for determination of matrix metalloproteinases: State of the art and recent advances. *J. Mater. Chem. B* **2020**. [CrossRef] [PubMed]

73. Vihinen, P.; Kähäri, V.-M. Matrix metalloproteinases in cancer: Prognostic markers and therapeutic targets. *Int. J. Cancer* **2002**, *99*, 157–166. [CrossRef] [PubMed]

74. Groblewska, M.; Siewko, M.; Mroczko, B.; Szmitkowski, M. The role of matrix metalloproteinases (MMPs) and their inhibitors (TIMPs) in the development of esophageal cancer. *Folia Histochem. Cytobiol.* **2012**, *50*, 12–19. [CrossRef] [PubMed]

75. Zhu, Y.-H.; Fu, L.; Chen, L.; Qin, Y.-R.; Liu, H.; Xie, F.; Zeng, T.; Dong, S.-S.; Li, J.; Li, Y.; et al. Downregulation of the Novel Tumor Suppressor DIRAS1 Predicts Poor Prognosis in Esophageal Squamous Cell Carcinoma. *Cancer Res.* **2013**, *73*, 2298–2309. [CrossRef]

76. Zou, S.; Yang, J.; Guo, J.; Su, Y.; He, C.; Wu, J.; Yu, L.; Ding, W.-Q.; Zhou, J. RAD18 promotes the migration and invasion of esophageal squamous cell cancer via the JNK-MMPs pathway. *Cancer Lett.* **2018**, *417*, 65–74. [CrossRef]

77. Zou, S.; Shang, Z.-F.; Liu, B.; Zhang, S.; Wu, J.; Huang, M.; Ding, W.-Q.; Zhou, J. DNA polymerase iota (Pol ι) promotes invasion and metastasis of esophageal squamous cell carcinoma. *Oncotarget* **2016**, *7*, 32274–32285. [CrossRef]

78. Garalla, H.M.; Lertkowit, N.; Tiszlavicz, L.; Reisz, Z.; Holmberg, C.; Beynon, R.; Simpson, D.; Varga, Á.; Kumar, J.D.; Dodd, S.; et al. Matrix metalloproteinase (MMP)-7 in Barrett's esophagus and esophageal adenocarcinoma: Expression, metabolism, and functional significance. *Physiol. Rep.* **2018**, *6*, e13683. [CrossRef]

79. Adachi, Y.; Itoh, F.; Yamamoto, H.; Matsuno, K.; Arimura, Y.; Kusano, M.; Endoh, T.; Hinoda, Y.; Oohara, M.; Hosokawa, M.; et al. Matrix metalloproteinase matrilysin (MMP-7) participates in the progression of human gastric and esophageal cancers. *Int. J. Oncol.* **1998**, *13*, 1031–1036. [CrossRef]

80. Wang, X.; Li, X.; Li, C.; He, C.; Ren, B.; Deng, Q.; Gao, W.; Wang, B. Aurora-A modulates MMP-2 expression via AKT/NF-κB pathway in esophageal squamous cell carcinoma cells. *Acta Biochim. Biophys. Sin.* **2016**, *48*, 520–527. [CrossRef]

81. Yamashita, K.; Mori, M.; Shiraishi, T.; Shibuta, K.; Sugimachi, K. Clinical significance of matrix metalloproteinase-7 expression in esophageal carcinoma. *Clin. Cancer Res.* **2000**, *6*, 1169–1174. [PubMed]

82. Yoshinaga, K.; Mimori, K.; Inoue, H.; Kamohara, Y.; Yamashita, K.; Tanaka, F.; Mori, M. Activin A enhances MMP-7 activity via the transcription factor AP-1 in an esophageal squamous cell carcinoma cell line. *Int. J. Oncol.* **2008**, *33*, 453–459. [CrossRef] [PubMed]

83. Wang, F.; Qi, Y.; Li, X.; He, W.; Fan, Q.-X.; Zong, H. HDAC inhibitor trichostatin A suppresses esophageal squamous cell carcinoma metastasis through HADC2 reduced MMP-2/9. *Clin. Investig. Med.* **2013**, *36*, 87–94. [CrossRef] [PubMed]

84. Wang, X.; Wang, J.; Jia, Y.; Wang, Y.; Han, X.; Duan, Y.; Lv, W.; Ma, M.; Liu, L. Methylation decreases the Bin1 tumor suppressor in ESCC and restoration by decitabine inhibits the epithelial mesenchymal transition. *Oncotarget* **2017**, *8*, 19661–19673. [CrossRef]

85. Beales, I.L.; Garcia-Morales, C.; Ogunwobi, O.O.; Mutungi, G. Adiponectin inhibits leptin-induced oncogenic signalling in oesophageal cancer cells by activation of PTP1B. *Mol. Cell. Endocrinol.* **2014**, *382*, 150–158. [CrossRef]

86. Allott, E.H.; Lysaght, J.; Cathcart, M.C.; Donohoe, C.L.; Cummins, R.; McGarrigle, S.A.; Kay, E.; Reynolds, J.V.; Pidgeon, G.P. MMP9 expression in oesophageal adenocarcinoma is upregulated with visceral obesity and is associated with poor tumour differentiation. *Mol. Carcinog.* **2013**, *52*, 144–154. [CrossRef]

87. Fan, F.; Jin, S.; Amundson, A.S.; Tong, T.; Fan, W.; Zhao, H.; Zhu, X.; Mazzacurati, L.; Li, X.; Petrik, K.L.; et al. ATF3 induction following DNA damage is regulated by distinct signaling pathways and over-expression of ATF3 protein suppresses cells growth. *Oncogene* **2002**, *21*, 7488–7496. [CrossRef]

88. Xie, J.-J.; Xie, Y.-M.; Chen, B.; Pan, F.; Guo, J.-C.; Zhao, Q.; Shen, J.-H.; Wu, Z.-Y.; Wu, J.-Y.; Xu, L.-Y.; et al. ATF3 functions as a novel tumor suppressor with prognostic significance in esophageal squamous cell carcinoma. *Oncotarget* **2014**, *5*, 8569–8582. [CrossRef]

89. Dasari, S.; Tchounwou, P.B. Cisplatin in cancer therapy: Molecular mechanisms of action. *Eur. J. Pharmacol.* **2014**, *740*, 364–378. [CrossRef]

90. Da Costa, N.M.; Pinto, L.F.R.; Nasciutti, L.E.; Palumbo, A., Jr. The Prominent Role of HMGA Proteins in the Early Management of Gastrointestinal Cancers. *BioMed Res. Int.* **2019**, *2019*, 2059516.

91. Lu, H.; Bhat, A.A.; Peng, D.; Chen, Z.; Zhu, S.; Hong, J.; Maacha, S.; Yan, J.; Robbins, D.J.; Washington, M.K.; et al. APE1 Upregulates MMP-14 via Redox-Sensitive ARF6-Mediated Recycling to Promote Cell Invasion of Esophageal Adenocarcinoma. *Cancer Res.* **2019**, *79*, 4426–4438. [CrossRef]

92. Da Costa, N.M.; Lima, S.C.S.; Simão, T.D.A.; Pinto, L.F.R. The potential of molecular markers to improve interventions through the natural history of oesophageal squamous cell carcinoma. *Biosci. Rep.* **2013**, *33*, 627–636. [CrossRef] [PubMed]

93. Wang, F.; He, W.; Fanghui, P.; Wang, L.; Fan, Q. NF-kBP65 promotes invasion and metastasis of oesophageal squamous cell cancer by regulating matrix metalloproteinase-9 and epithelial-to-mesenchymal transition. *Cell Biol. Int.* **2013**, *37*, 780–788. [CrossRef] [PubMed]

94. Shin, W.-S.; Hong, Y.; Lee, H.W.; Lee, S.-T. Catalytically defective receptor protein tyrosine kinase PTK7 enhances invasive phenotype by inducing MMP-9 through activation of AP-1 and NF-κB in esophageal squamous cell carcinoma cells. *Oncotarget* **2016**, *7*, 73242–73256. [CrossRef]

95. Liu, N.; Zhang, R.; Wu, J.; Pu, Y.; Yin, X.; Cheng, Y.; Wu, J.; Feng, C.; Luo, Y.; Zhang, J. Interleukin-17A promotes esophageal adenocarcinoma cell invasiveness through ROS-dependent, NF-κB-mediated MMP-2/9 activation. *Oncol. Rep.* **2017**, *37*, 1779–1785. [CrossRef]

96. Clemons, N.J.; Shannon, N.B.; Abeyratne, L.R.; Walker, C.E.; Saadi, A.; O'Donovan, M.L.; Lao-Sirieix, P.P.; Fitzgerald, R.C. Nitric oxide-mediated invasion in Barrett's high-grade dysplasia and adenocarcinoma. *Carcinogenesis* **2010**, *31*, 1669–1675. [CrossRef]

97. Murray, G.I.; Duncan, M.E.; O'Neil, P.; McKay, J.A.; Melvin, W.T.; Fothergill, J.E. Matrix metalloproteinase-1 is associated with poor prognosis in oesophageal cancer. *J. Pathol.* **1998**, *185*, 256–261. [CrossRef]

98. Yamashita, K.; Mori, M.; Kataoka, A.; Inoue, H.; Sugimachi, K. The clinical significance of MMP-1 expression in oesophageal carcinoma. *Br. J. Cancer* **2001**, *84*, 276–282. [CrossRef]

99. Cheung, W.Y.; Zhai, R.; Bradbury, P.; Hopkins, J.; Kulke, M.H.; Heist, R.S.; Asomaning, K.; Ma, C.; Xu, W.; Wang, Z.; et al. Single nucleotide polymorphisms in the matrix metalloproteinase gene family and the frequency and duration of gastroesophageal reflux disease influence the risk of esophageal adenocarcinoma. *Int. J. Cancer* **2012**, *131*, 2478–2486. [CrossRef]

100. Bradbury, P.A.; Zhai, R.; Hopkins, J.; Kulke, M.H.; Heist, R.S.; Singh, S.; Zhou, W.; Ma, C.; Xu, W.; Asomaning, K.; et al. Matrix metalloproteinase 1, 3 and 12 polymorphisms and esophageal adenocarcinoma risk and prognosis. *Carcinogenesis* **2009**, *30*, 793–798. [CrossRef]

101. Peng, B.; Cao, L.; Ma, X.; Wang, W.; Wang, D.; Yu, L. Meta-analysis of association between matrix metalloproteinases 2,7 and 9 promoter polymorphisms and cancer risk. *Mutagenesis* **2010**, *25*, 371–379. [CrossRef]

102. Petty, R.D.; Dahle-Smith, A.; Stevenson, D.A.; Osborne, A.; Massie, D.; Clark, C.; Murray, G.I.; Dutton, S.J.; Roberts, C.; Chong, I.Y.; et al. Gefitinib and EGFR Gene Copy Number Aberrations in Esophageal Cancer. *J. Clin. Oncol.* **2017**, *35*, 2279–2287. [CrossRef] [PubMed]

103. Shima, I.; Sasaguri, Y.; Arima, N.; Yamana, H.; Fujita, H.; Morimatsu, M.; Nagase, H. Expression of epidermal growth-factor (EGF), matrix metalloproteinase-9 (MMP-9) and proliferating cell nuclear antigen (PCNA) in esophageal cancer. *Int. J. Oncol.* **1995**, *6*, 833–839. [CrossRef] [PubMed]

104. Shima, I.; Sasaguri, Y.; Kusukawa, J.; Nakano, R.; Yamana, H.; Fujita, H.; Kakegawa, T.; Morimatsu, M. Production of matrix metalloproteinase 9 (92-kDa gelatinase) by human oesophageal squamous cell carcinoma in response to epidermal growth factor. *Br. J. Cancer* **1993**, *67*, 721–727. [CrossRef]

105. Okawa, T.; Michaylira, C.Z.; Kalabis, J.; Stairs, U.B.; Nakagawa, H.; Andl, C.; Johnstone, C.N.; Klein-Szanto, A.J.; El-Deiry, W.S.; Cukierman, E.; et al. The functional interplay between EGFR overexpression, hTERT activation, and p53 mutation in esophageal epithelial cells with activation of stromal fibroblasts induces tumor development, invasion, and differentiation. *Genome Res.* **2007**, *21*, 2788–2803. [CrossRef]

106. Ellerbroek, S.M.; Halbleib, J.M.; Benavidez, M.; Warmka, J.K.; Wattenberg, E.V.; Stack, M.S.; Hudson, L.G. Phosphatidylinositol 3-kinase activity in epidermal growth factor-stimulated matrix metalloproteinase-9 production and cell surface association. *Cancer Res.* **2001**, *61*, 1855–1861.

107. Maziveyi, M.; Alahari, S.K. Cell matrix adhesions in cancer: The proteins that form the glue. *Oncotarget* **2017**, *8*, 48471–48487. [CrossRef]

108. Shams, H.; Hoffman, B.D.; Mofrad, M.R.K. The "Stressful" Life of Cell Adhesion Molecules: On the Mechanosensitivity of Integrin Adhesome. *J. Biomech. Eng.* **2018**, *140*, 020807. [CrossRef]

109. Canel, M.; Serrels, A.; Frame, M.C.; Brunton, V.G.; Frisch, S.M.; Schaller, M.; Cieply, B. E-cadherin-integrin crosstalk in cancer invasion and metastasis. *J. Cell Sci.* **2013**, *126*, 393–401. [CrossRef]

110. Liu, B.; Li, X.; Li, C.; Xu, R.; Sun, X. miR-25 mediates metastasis and epithelial–mesenchymal-transition in human esophageal squamous cell carcinoma via regulation of E-cadherin signaling. *Bioengineered* **2019**, *10*, 679–688. [CrossRef]

111. Xu, X.L.; Ling, Z.Q.; Chen, S.Z.; Li, B.; Ji, W.H.; Mao, W.M. The impact of E-cadherin expression on the prognosis of esophageal cancer: A meta-analysis. *Dis Esophagus* **2014**, *27*, 79–86. [CrossRef]

112. Lin, Y.; Shen, L.Y.; Fu, H.; Dong, B.; Yang, H.L.; Yan, W.P.; Kang, X.Z.; Dai, L.; Zhou, H.T.; Yang, Y.B.; et al. P21, COX-2, and E-cadherin are potential prognostic factors for esophageal squamous cell carcinoma. *Dis. Esophagus* **2017**, *30*, 1–10. [CrossRef] [PubMed]

113. Kim, W.; Kang, Y.S.; Kim, J.S.; Shin, N.Y.; Hanks, S.K.; Song, W.K. The integrin-coupled signaling adaptor p130Cas suppresses Smad3 function in transforming growth factor-beta signaling. *Mol. Biol. Cell* **2008**, *19*, 2135–2146. [CrossRef]

114. Fu, L.; Qin, Y.R.; Xie, D.; Chow, H.Y.; Ngai, S.M.; Kwong, R.L.W.; Li, Y.; Guan, X.Y. Identification of alpha-actinin 4 and 67 kDa laminin receptor as stage-specific markers in esophageal cancer via proteomic approaches. *Cancer* **2007**, *110*, 2672–2681. [CrossRef]

115. Zhang, J.; Zhi, H.; Zhou, C.; Ding, F.; Luo, A.; Zhang, X.; Sun, Y.; Wang, X.; Wu, M.; Liu, Z. Up-regulation of fibronectin in oesophageal squamous cell carcinoma is associated with activation of the Erk pathway. *J. Pathol.* **2005**, *207*, 402–409. [CrossRef] [PubMed]

116. Durbeej, M. Laminins. *Cell Tissue Res.* **2010**, *339*, 259–268. [CrossRef] [PubMed]

117. Takubo, K.; Tanaka, Y.; Miyama, T.; Fujita, K.; Mafune, K.-I. Basaloid-Squamous Carcinoma of the Esophagus with Marked Deposition of Basement Membrane Substance. *Pathol. Int.* **1991**, *41*, 59–64. [CrossRef]

118. Mori, M.; Shimono, R.; Kido, A.; Kuwano, H.; Akazawa, K.; Sugimachi, K. Distribution of basement membrane antigens in human esophageal lesions: An immunohistochemical study. *Int. J. Cancer* **1991**, *47*, 839–842. [CrossRef]

119. Baba, K.; Kuwano, H.; Kitamura, K.; Sugimachi, K. Carcinomatous invasion and lymphocyte infiltration in early esophageal carcinoma with special regard to the basement membrane. An immunohistochemical study. *Hepatogastroenterology* **1993**, *40*, 226–231.

120. Yamamoto, H.; Itoh, F.; Iku, S.; Hosokawa, M.; Imai, K. Expression of the gamma(2) chain of laminin-5 at the invasive front is associated with recurrence and poor prognosis in human esophageal squamous cell carcinoma. *Clin. Cancer Res.* **2001**, *7*, 896–900.

121. Fukai, Y.; Masuda, N.; Kato, H.; Fukuchi, M.; Miyazaki, T.; Nakajima, M.; Sohda, M.; Kuwano, H.; Nakajima, T. Correlation between laminin-5 gamma2 chain and epidermal growth factor receptor expression in esophageal squamous cell carcinomas. *Oncology* **2005**, *69*, 71–80. [CrossRef] [PubMed]

122. Mizushima, H.; Miyagi, Y.; Kikkawa, Y.; Yamanaka, N.; Yasumitsu, H.; Misugi, K.; Miyazaki, K. Differential Expression of Laminin-5/Ladsin Subunits in Human Tissues and Cancer Cell Lines and Their Induction by Tumor Promoter and Growth Factors. *J. Biochem.* **1996**, *120*, 1196–1202. [CrossRef] [PubMed]

123. Schenk, S.; Hintermann, E.; Bilban, M.; Koshikawa, N.; Hojilla, C.; Khokha, R.; Quaranta, V. Binding to EGF receptor of a laminin-5 EGF-like fragment liberated during MMP-dependent mammary gland involution. *J. Cell Biol.* **2003**, *161*, 197–209. [CrossRef] [PubMed]

124. Shen, X.M.; Wu, Y.P.; Feng, Y.B.; Luo, M.L.; Du, X.L.; Zhang, Y.; Cai, Y.; Xu, X.; Han, Y.L.; Zhang, X.; et al. Interaction of MT1-MMP and laminin-5gamma2 chain correlates with metastasis and invasiveness in human esophageal squamous cell carcinoma. *Clin. Exp. Metastasis* **2007**, *24*, 541–550. [CrossRef] [PubMed]

125. Kita, Y.; Mimori, K.; Tanaka, F.; Matsumoto, T.; Haraguchi, N.; Ishikawa, K.; Matsuzaki, S.; Fukuyoshi, Y.; Inoue, H.; Natsugoe, S.; et al. Clinical significance of LAMB3 and COL7A1 mRNA in esophageal squamous cell carcinoma. *Eur. J. Surg. Oncol. EJSO* **2009**, *35*, 52–58. [CrossRef]

126. Baba, Y.; Iyama, K.-I.; Hirashima, K.; Nagai, Y.; Yoshida, N.; Hayashi, N.; Miyanari, N.; Baba, H. Laminin-332 promotes the invasion of oesophageal squamous cell carcinoma via PI3K activation. *Br. J. Cancer* **2008**, *98*, 974–980. [CrossRef]

127. Aghcheli, K.; Parsian, H.; Qujeq, D.; Talebi, M.; Mosapour, A.; Khalilipour, E.; Islami, F.; Semnani, S.; Malekzadeh, R. Serum hyaluronic acid and laminin as potential tumor markers for upper gastrointestinal cancers. *Eur. J. Intern. Med.* **2012**, *23*, 58–64. [CrossRef]

128. Lin, J.; Myers, A.L.; Wang, Z.; Nancarrow, D.J.; Ferrer-Torres, D.; Handlogten, A.; Leverenz, K.; Bao, J.; Thomas, D.G.; Wang, T.D.; et al. Osteopontin (OPN/SPP1) isoforms collectively enhance tumor cell invasion and dissemination in esophageal adenocarcinoma. *Oncotarget* **2015**, *6*, 22239–22257. [CrossRef]

129. Meng, X.; Chen, X.; Lu, P.; Ma, W.; Yue, D.; Song, L.; Fan, Q. MicroRNA-202 inhibits tumor progression by targeting LAMA1 in esophageal squamous cell carcinoma. *Biochem. Biophys. Res. Commun.* **2016**, *473*, 821–827. [CrossRef]

130. Yang, Y.; Li, D.; Yang, Y.; Jiang, G. An integrated analysis of the effects of microRNA and mRNA on esophageal squamous cell carcinoma. *Mol. Med. Rep.* **2015**, *12*, 945–952. [CrossRef]

131. Ma, G.; Zhang, F.; Dong, X.; Wang, X.; Ren, Y. Low expression of microRNA-202 is associated with the metastasis of esophageal squamous cell carcinoma. *Exp. Ther. Med.* **2016**, *11*, 951–956. [CrossRef] [PubMed]

132. Sudo, T.; Iwaya, T.; Nishida, N.; Sawada, G.; Takahashi, Y.; Ishibashi, M.; Shibata, K.; Fujita, H.; Shirouzu, K.; Mori, M.; et al. Expression of mesenchymal markers vimentin and fibronectin: The clinical significance in esophageal squamous cell carcinoma. *Ann. Surg. Oncol.* **2013**, *20*, S324–S335. [CrossRef] [PubMed]

133. Xiao, J.; Yang, W.; Xu, B.; Zhu, H.; Zou, J.; Su, C.; Rong, J.; Wang, T.; Chen, Z. Expression of fibronectin in esophageal squamous cell carcinoma and its role in migration. *BMC Cancer* **2018**, *18*, 976. [CrossRef] [PubMed]

134. Rousseau, M.C.; Hsu, R.Y.; Spicer, J.D.; McDonald, B.; Chan, C.H.; Perera, R.M.; Giannias, B.; Chow, S.C.; Rousseau, S.; Law, S.; et al. Lipopolysaccharide-induced toll-like receptor 4 signaling enhances the migratory ability of human esophageal cancer cells in a selectin-dependent manner. *Surgery* **2013**, *154*, 69–77. [CrossRef] [PubMed]

135. Poehlmann, A.; Kuester, D.; Malfertheiner, P.; Guenther, T.; Roessner, A. Inflammation and Barrett's carcinogenesis. *Pathol. Res. Pract.* **2012**, *208*, 269–280. [CrossRef] [PubMed]

136. He, F.; Ai, B.; Tian, L. Identification of genes and pathways in esophageal adenocarcinoma using bioinformatics analysis. *Biomed. Rep.* **2018**, *9*, 305–312. [CrossRef]

137. Leppänen, J.; Bogdanoff, S.; Lehenkari, P.P.; Saarnio, J.; Kauppila, J.H.; Karttunen, T.J.; Huhta, H.; Helminen, O. Tenascin-C and fibronectin in normal esophageal mucosa, Barrett's esophagus, dysplasia and adenocarcinoma. *Oncotarget* **2017**, *8*, 66865–66877. [CrossRef]

138. Kuo, I.-Y.; Wu, C.-C.; Chang, J.-M.; Huang, Y.-L.; Lin, C.-H.; Yan, J.-J.; Sheu, B.-S.; Lu, P.-J.; Chang, W.-L.; Lai, W.-W.; et al. Low SOX17 expression is a prognostic factor and drives transcriptional dysregulation and esophageal cancer progression. *Int. J. Cancer* **2014**, *135*, 563–573. [CrossRef]

139. Wei, Q.; Li, X.; Yu, W.; Zhao, K.; Qin, G.; Chen, H.; Gu, Y.; Ding, F.; Zhu, Z.; Fu, X.; et al. microRNA-messenger RNA regulatory network of esophageal squamous cell carcinoma and the identification of miR-1 as a biomarker of patient survival. *J. Cell. Biochem.* **2019**, *120*, 12259–12272. [CrossRef]

140. Ma, J.; Xiao, Y.; Tian, B.; Chen, S.; Zhang, B.; Wu, J.; Wu, Z.; Li, X.; Tang, J.; Yang, D.; et al. Long noncoding RNA lnc-ABCA12-3 promotes cell migration, invasion, and proliferation by regulating fibronectin 1 in esophageal squamous cell carcinoma. *J. Cell. Biochem.* **2019**, *121*, 1374–1387. [CrossRef]

141. Zhang, C.-Y.; Li, R.-K.; Qi, Y.; Yang, Y.; Liu, D.-L.; Zhao, J.; Zhu, D.-Y.; Wu, K.; Zhou, X.-D. Upregulation of long noncoding RNA SPRY4-IT1 promotes metastasis of esophageal squamous cell carcinoma via induction of epithelial–mesenchymal transition. *Cell Biol. Toxicol.* **2016**, *32*, 391–401. [CrossRef] [PubMed]

142. Wu, X.; Qiao, B.; Liu, Q.; Zhang, W. Upregulation of extracellular matrix metalloproteinase inducer promotes hypoxia-induced epithelial-mesenchymal transition in esophageal cancer. *Mol. Med. Rep.* **2015**, *12*, 7419–7424. [CrossRef] [PubMed]

143. Broll, R.; Meyer, S.; Neuber, M.; Bruch, H.P. Expression of tenascin in tumors of the esophagus, small intestine and colorectum. An immunohistochemical study. *Gen. Diagn. Pathol.* **1995**, *141*, 111–119.

144. Salmela, M.T.; Karjalainen-Lindsberg, M.L.; Puolakkainen, P.; Saarialho-Kere, U. Upregulation and differential expression of matrilysin (MMP-7) and metalloelastase (MMP-12) and their inhibitors TIMP-1 and TIMP-3 in Barrett's oesophageal adenocarcinoma. *Br. J. Cancer* **2001**, *85*, 383–392. [CrossRef]

145. Ohtsuka, M.; Yamamoto, H.; Oshiro, R.; Takahashi, H.; Masuzawa, T.; Uemura, M.; Haraguchi, N.; Nishimura, J.; Hata, T.; Yamasaki, M.; et al. Concurrent expression of C4.4A and Tenascin-C in tumor cells relates to poor prognosis of esophageal squamous cell carcinoma. *Int. J. Oncol.* **2013**, *43*, 439–446. [CrossRef] [PubMed]

146. Yang, Z.T.; Yeo, S.Y.; Yin, Y.X.; Lin, Z.H.; Lee, H.M.; Xuan, Y.H.; Cui, Y.; Kim, S.H. Tenascin-C, a prognostic determinant of esophageal squamous cell carcinoma. *PLoS ONE* **2016**, *11*, e0145807. [CrossRef]

147. Yang, Z.; Zhang, C.; Feng, Y.; Qi, W.; Cui, Y.; Xuan, Y. Tenascin-C is involved in promotion of cancer stemness via the Akt/HIF1α axis in esophageal squamous cell carcinoma. *Exp. Mol. Pathol.* **2019**, *109*, 104239. [CrossRef]

148. Fortuna-Costa, A.; Gomes, A.M.; Kozlowski, E.O.; Stelling, M.P.; Pavão, M.S.G. Extracellular Galectin-3 in Tumor Progression and Metastasis. *Front. Oncol.* **2014**, *4*, 138. [CrossRef]

149. Cardoso, A.C.F.; Andrade, L.N.D.S.; Bustos, S.O.; Chammas, R. Galectin-3 Determines Tumor Cell Adaptive Strategies in Stressed Tumor Microenvironments. *Front. Oncol.* **2016**, *6*, 740. [CrossRef]

150. Kashyap, M.K.; Harsha, H.C.; Renuse, S.; Pawar, H.; Sahasrabuddhe, N.A.; Kim, M.-S.; Marimuthu, A.; Keerthikumar, S.; Muthusamy, B.; Kandasamy, K.; et al. SILAC-based quantitative proteomic approach to identify potential biomarkers from the esophageal squamous cell carcinoma secretome. *Cancer Biol. Ther.* **2010**, *10*, 796–810. [CrossRef]

151. Çobanoğlu, U.; Mergan, D.; Dulger, A.C.; Celik, S.; Kemik, O.; Sayir, F. Are Serum Mac 2-Binding Protein Levels Elevated in Esophageal Cancer? A Control Study of Esophageal Squamous Cell Carcinoma Patients. *Dis. Markers* **2018**, *2018*, 3610239. [CrossRef] [PubMed]

152. Qiao, L.; Liang, N.; Xie, J.; Luo, H.; Zhang, J.; Deng, G.; Li, Y.; Zhang, J. Gene silencing of galectin-3 changes the biological behavior of Eca109 human esophageal cancer cells. *Mol. Med. Rep.* **2016**, *13*, 160–166. [CrossRef] [PubMed]

153. Zhang, J.; Deng, G.; Qiao, L.; Luo, H.; Liu, O.; Liang, N.; Xie, J.; Zhang, J. Effect of galectin-3 on vasculogenic mimicry in esophageal cancer cells. *Oncol. Lett.* **2018**, *15*, 4907–4911. [CrossRef] [PubMed]

154. Cui, G.; Cui, M.; Li, Y.; Liang, Y.; Li, W.; Guo, H.; Zhao, S. Galectin-3 knockdown increases gefitinib sensitivity to the inhibition of EGFR endocytosis in gefitinib-insensitive esophageal squamous cancer cells. *Med. Oncol.* **2015**, *32*, 124. [CrossRef]

155. Nagahara, K.; Arikawa, T.; Oomizu, S.; Kontani, K.; Nobumoto, A.; Tateno, H.; Watanabe, K.; Niki, T.; Katoh, S.; Miyake, M.; et al. Galectin-9 increases Tim-3+ dendritic cells and CD8+ T cells and enhances antitumor immunity via galectin-9-Tim-3 interactions. *J. Immunol.* **2008**, *181*, 7660–7669. [CrossRef]

156. Chiyo, T.; Fujita, K.; Iwama, H.; Fujihara, S.; Tadokoro, T.; Ohura, K.; Matsui, T.; Goda, Y.; Kobayashi, N.; Nishiyama, N.; et al. Galectin-9 Induces Mitochondria-Mediated Apoptosis of Esophageal Cancer In Vitro and In Vivo in a Xenograft Mouse Model. *Int. J. Mol. Sci.* **2019**, *20*, 2634. [CrossRef]

157. Akashi, E.; Fujihara, S.; Morishita, A.; Tadokoro, T.; Chiyo, T.; Fujikawa, K.; Kobara, H.; Mori, H.; Iwama, H.; Okano, K.; et al. Effects of galectin-9 on apoptosis, cell cycle and autophagy in human esophageal adenocarcinoma cells. *Oncol. Rep.* **2017**, *38*, 506–514. [CrossRef]

158. Zhu, X.; Ding, M.; Yu, M.-L.; Feng, M.-X.; Tan, L.-J.; Zhao, F.-K. Identification of galectin-7 as a potential biomarker for esophageal squamous cell carcinoma by proteomic analysis. *BMC Cancer* **2010**, *10*, 290. [CrossRef]

159. Twarock, S.; Freudenberger, T.; Poscher, E.; Dai, G.; Jannasch, K.; Dullin, C.; Alves, F.; Prenzel, K.; Knoefel, W.T.; Stoecklein, N.H.; et al. Inhibition of Oesophageal Squamous Cell Carcinoma Progression by in vivo Targeting of Hyaluronan Synthesis. *Mol. Cancer* **2011**, *10*, 30. [CrossRef]
160. Thelin, M.A.; Svensson, K.J.; Shi, X.; Bagher, M.; Axelsson, J.; Isinger-Ekstrand, A.; Van Kuppevelt, T.H.; Johansson, J.; Nilbert, M.; Zaia, J.; et al. Dermatan sulfate is involved in the tumorigenic properties of esophagus squamous cell carcinoma. *Cancer Res.* **2012**, *72*, 1943–1952. [CrossRef]
161. Twarock, S.; Tammi, M.I.; Savani, R.C.; Fischer, J.W. Hyaluronan Stabilizes Focal Adhesions, Filopodia, and the Proliferative Phenotype in Esophageal Squamous Carcinoma Cells. *J. Biol. Chem.* **2010**, *285*, 23276–23284. [CrossRef] [PubMed]

Article

CCN-Based Therapeutic Peptides Modify Pancreatic Ductal Adenocarcinoma Microenvironment and Decrease Tumor Growth in Combination with Chemotherapy

Andrea Resovi [1], Patrizia Borsotti [1], Tommaso Ceruti [2], Alice Passoni [3], Massimo Zucchetti [2], Alexander Berndt [4], Bruce L. Riser [5,6], Giulia Taraboletti [1,*] and Dorina Belotti [1,*]

[1] Laboratory of Tumor Microenvironment, Department of Oncology, Istituto di Ricerche Farmacologiche Mario Negri IRCCS, 24126 Bergamo, Italy; andrea.resovi@marionegri.it (A.R.); patrizia.borsotti@marionegri.it (P.B.)
[2] Laboratory of Cancer Pharmacology, Department of Oncology, Istituto di Ricerche Farmacologiche Mario Negri IRCCS, 20156 Milan, Italy; tommaso.ceruti@marionegri.it (T.C.); massimo.zucchetti@marionegri.it (M.Z.)
[3] Laboratory of Mass Spectrometry, Department of Environmental Health Sciences, Istituto di Ricerche Farmacologiche Mario Negri IRCCS, 20156 Milan, Italy; alice.passoni@marionegri.it
[4] Section Pathology, Institute of Legal Medicine, Jena University Hospital, D-07747 Jena, Germany; Alexander.Berndt@med.uni-jena.de
[5] BLR Bio LLC, Kenosha, WI 53140, USA; riser@blrbio.com
[6] Department of Physiology and Biophysics, and Department of Medicine Center for Cancer Cell Biology, Immunology and Infection, Rosalind Franklin University of Medicine and Science, North Chicago, IL 60064, USA
* Correspondence: giulia.taraboletti@marionegri.it (G.T.); dorina.belotti@marionegri.it (D.B.)

Received: 6 February 2020; Accepted: 9 April 2020; Published: 13 April 2020

Abstract: The prominent desmoplastic stroma of pancreatic ductal adenocarcinoma (PDAC) is a determinant factor in tumor progression and a major barrier to the access of chemotherapy. The PDAC microenvironment therefore appears to be a promising therapeutic target. CCN2/CTGF is a profibrotic matricellular protein, highly present in the PDAC microenvironment and associated with disease progression. Here we have investigated the therapeutic value of the CCN2-targeting BLR100 and BLR200, two modified synthetic peptides derived from active regions of CCN3, an endogenous inhibitor of CCN2. In a murine orthotopic PDAC model, the two peptides, administered as monotherapy at low doses (approximating physiological levels of CCN3), had tumor inhibitory activity that increased with the dose. The peptides affected the tumor microenvironment, inhibiting fibrosis and vessel formation and reducing necrosis. Both peptides were active in preventing ascites formation. An increased activity was obtained in combination regimens, administering BLR100 or BLR200 with the chemotherapeutic drug gemcitabine. Pharmacokinetic analysis indicated that the improved activity of the combination was not mainly determined by the substantial increase in gemcitabine delivery to tumors, suggesting other effects on the tumor microenvironment. The beneficial remodeling of the tumor stroma supports the potential value of these CCN3-derived peptides for targeting pathways regulated by CCN2 in PDAC.

Keywords: PDAC; tumor microenvironment; CCN2/CCN3; matricellular proteins

---

1. Introduction

Pancreatic ductal adenocarcinoma (PDAC) is one of the most aggressive tumor types, with limited treatment options. It is characterized by a strong desmoplastic reaction, with prominent fibrosis and extracellular matrix (ECM) deposition that, on the one hand, promote PDAC progression, influencing tumor cell proliferation, survival, and invasion and, on the other, hampers distribution and activity of chemotherapy and immunotherapy [1]. This has generated great interest for the development of agents targeting the tumor microenvironment as new therapeutic tools for PDAC treatment, particularly when used in combination with cytotoxic drugs [2]. Attempts to directly deplete PDAC associated stroma cell populations often resulted in a more aggressive tumor [3,4]. However, targeting the molecular mechanisms governing the desmoplastic reaction to affect the tumor promoting activity of the microenvironment emerged as a more promising therapeutic approach [2,5].

In this context, the unique functional properties of CCN matricellular proteins place them in an area of great interest for the design of new drugs that mimic or regulate their biological activity. The CCNs are considered to be master regulators of multiple biological functions. They are produced and secreted by a variety of cell types including many tumors [6] and are found circulating in plasma or deposited in tissues where they simultaneously bind to a variety of ECM molecules and cell receptors, translating environmental signals to the cell. The four modules that make up all but one of the six-member family (CCN5 contains only three modules) contain sites for binding different receptors. These receptors include Notch, integrins, TrkA, and heparan sulfate proteoglycans, and they are known to modulate such diverse activities as cell replication, death, adhesion, motility, and ECM production [7].

Among the CCNs, CCN2 (formerly named connective tissue growth factor, CTGF) is known to play a role in early development of a variety of tissues, but mostly in pathological conditions where it interacts with ECM molecules, growth factors, and cell receptors depending on the cellular context. CCN2 is a potent inducer of ECM production [8] and a profibrotic agent in pancreatic disease [9–11], renal pathology [12,13], bronchopulmonary dysplasia [14], and cardiac fibrosis [15]. Its altered expression has been associated with tumorigenesis [16].

In particular, CCN2 acts as a profibrotic molecule in PDAC [17], where it is produced by acinar cells, ductal cells, stellate cells, and fibroblasts. CCN2 plays an important role in the activation of pancreatic stellate cells, stimulating the production of ECM, proteases, and protease inhibitors and modulating the activity of growth factors and angiogenic factors. Activated stellate cells, in turn, are the major source of CCN2 during PDAC progression. We have recently demonstrated that plasma CCN2 is upregulated in PDAC patients and is a valuable potential biomarker for PDAC diagnosis and for monitoring drug response [18].

CCN2 has recently been proposed as a target for PDAC therapy. The CCN2-targeting mAb FG-3019 inhibited tumor growth and metastasis in the PANC-1 orthotopic preclinical model [19]. In genetically engineered mice *LSL-Kras*$^{G12D/+}$*;LSL-Trp53*$^{R172H/+}$*;Pdx-1-Cre* (KPC), FG-3019 enhanced tumor response to the chemotherapeutic drug gemcitabine [20]. FG-3019 (pamrevlumab) is currently in a Phase 3 clinical trial to evaluate its efficacy and safety as neoadjuvant treatment in combination with gemcitabine plus nab-paclitaxel in the treatment of locally advanced, unresectable pancreatic cancer.

CCN3/NOV, another member of the CCN family, acts as an endogenous inhibitor of CCN2 biological activity as well as production [21]. In particular, CCN3 inhibited CCN2 profibrotic activity in in vitro and in vivo models of renal disease, where it blocked cellular injury and prevented the conversion of mesangial cells to activated alpha-smooth muscle actin-positive fibroblast-like cells [22,23]. CCN3 also reduced skin fibrosis blocking collagen type 1 production and cell proliferation stimulated by platelet derived growth factor (PDGF) [24]. Building on the endogenous regulatory role of CCN3 on CCN2, a set of small modified peptides based on two CCN3 regions identified as responsible for this activity have been created by BLR Bio, as potential agents to treat fibrotic diseases including cancer [24].

Given the role of CCN2 in PDAC progression, this study was designed to investigate the potential value of two of these peptides, BLR100 and BLR200, for use as therapeutic agents to treat PDAC, using a PDAC model transplanted orthotopically in the pancreas of immunocompetent mice. The activity of the peptides has been evaluated in tumors formed by the FC1199 tumor cells, derived from KPC mice, the model used to first demonstrate the activity of the CCN2 inhibitor FG-3019 in PDAC. When implanted orthotopically in syngeneic mice, these cells form tumors that recapitulate the pathological features of the original tumors in genetically engineered mouse model (GEMM), as well as of human PDAC, particularly in terms of desmoplastic microenvironment. This study shows the ability of BLR100 and BLR200 to modify the PDAC microenvironment, controlling tumor growth and ascites formation, particularly in combination with chemotherapy.

2. Materials and Methods

2.1. Drugs

BLR100 and BLR200 and the control peptide (scrambled) were chemically synthesized by JPT Peptide Technologies (Berlin, Germany). BLR100 and BLR200 are based on two different 14 amino acid sequences identified from different modules in CCN3, selected for ability to interact with CCN2 and block CCN2 binding to cell receptors (Riser, B.L., Inventor, "CCN3 peptides and analogs thereof for therapeutic use." U.S. Patent 8518,395, Issued 27 August 2013) with proprietary modifications to increase stability. The purity of the peptides used were greater than 95%, with the remaining 5% representing small unincorporated amino acid groups, as determined by HPLC analysis.

The peptides were dissolved in water at the concentration of 1 mM, stored at −80 °C, and further diluted in phosphate buffered saline immediately before use. Gemcitabine (Teva, Assago, Italy) was dissolved in saline (40 mg/mL), stored at −80 °C, and further diluted immediately before use.

2.2. Tumor Cells

The FC1199 pancreatic cancer cell line, derived from tumors arisen in *LSL-Kras$^{G12D/+}$;LSL-Trp53$^{R172H/+}$;Pdx-1-Cre* mice [25] in the C57BL/6 background, was provided by D.A. Tuveson (Cold Spring Harbor, NY, USA). Cells were cultured in Dulbecco modified Eagle's medium (DMEM) (Gibco, ThermoFisher Scientific, Rodano, Italy) supplemented with 10% Fetal calf serum (FCS) (Euroclone, Milano, Italy) and 1% L-glutamine (Gibco). Cells were kept in culture for no more than three weeks before injection in mice and routinely tested and found free of mycoplasma infection.

2.3. Proliferation Assay

FC1199 were seeded into 96-well plates in DMEM supplemented with 5% FCS. At 24 h after seeding, cells were exposed to increasing concentrations of BLR100 and BLR200 alone or in combination with increasing concentrations of gemcitabine. After a 72 h incubation, cells were fixed and stained with crystal violet solution (Sigma–Aldrich, Merck Life Science, Milano, Italy). The staining was eluted with a 1:1 ethanol/0.1 M sodium citrate solution, and the absorbance at 595 nm was measured. Each condition was tested in triplicate.

2.4. ELISA Assays

CCN2 levels in conditioned media and cell lysate of FC1199 tumor cells were measured by ELISA (CSB-E07877m, Cusabio, Wuhan, China), according to the manufacturer's instructions. Each sample was analyzed in duplicate.

2.5. In Vivo Studies

Procedures involving animals and their care were conducted in conformity with institutional guidelines that comply with national (Lgs 26/2014) and EU directives laws and policies (EEC Council Directive 2010/63) in line with guidelines for the welfare and use of animals in cancer research [26] and with the "3Rs" principle. Animal studies were approved by the Mario Negri Institute Animal Care and Use Committee and by the Italian Ministry of Health (Authorization 125/2016-PR).

Six- to eight-week old female C57BL/6 mice (Charles River Laboratories, Lecco, Italy) were maintained under specific-pathogen-free conditions, with constant temperature and humidity, and handled using aseptic procedures. FC1199 cells (5×10^4) were implanted orthotopically in the pancreas as described [18]. Peptides were administered i.p. at the dose of 4.5 or 10 μg/kg, as indicated, with the schedule shown in Figure 2A and Figure 4A. Gemcitabine was administered i.v. at 40 mg/kg (see scheme in Figure 4A). Control groups received the same volume of vehicle. Mice were weighed every other day as a measure of drug toxicity. When the first mice displayed signs of distress the experiment was concluded for all experimental groups. Tumor burden was evaluated at the end of the experiment, as pancreas weight. Peritoneal ascitic fluid was collected with a syringe and the volume recorded. Animals with a collectable peritoneal fluid ($\geq$100 μL) were considered positive for ascites. Liver of mice were analyzed for macroscopic metastases, and no metastatic foci was detected in any group.

2.6. Pharmacokinetics Studies

FC1199-bearing mice were treated i.p with BLR200 (10 μg/kg) from day 11 for a total of 4 treatments, every other day. On day 17, mice were treated with gemcitabine (100 mg/kg, i.v., single bolus). After 30 min and 1 h, plasma and tumors were collected for HPLC measurement of gemcitabine and metabolites. The plasma fraction added with 10 μL tetrahydrouridine (THU) 2.5 mg/mL was immediately separated by centrifugation at 3200$\times$ *g* for 15 min at 4 °C and stored at −20 °C until analysis. Tumors were immediately frozen in liquid nitrogen and stored at −20 °C until analysis. Extraction and analysis of gemcitabine (dFdc, dFdCTP, and dFdU) in tumors and plasma were carried out according to Bapiro et al. [27].

Briefly, tumors were weighted, spiked with 2′-deossicitidine as the internal standard (IS) at a final concentration of 1 ng/mg, and homogenized with ice-cold acetonitrile (ACN) 50% (v/v) containing THU 25 μg/mL in a Precellys Evolution homogenizer (Bertin Technologies S.A.S., Montigny-le-Bretonneux, France). Fifty microliters of homogenate were added with 200 μL ACN 50% (v/v), vortexed, and centrifuged at 13,200 rpm for 10 min at 4 °C. The supernatant was dried under N_2 flux and the residue reconstituted with 100 μL of milliQ water. Concerning plasma, 25 μL were spiked with 1 μL of 10 μg/mL of IS and processed in the same way as tumors. The reconstituted tumors and plasma extracts were analyzed by injecting 2 μL in a UHPLC System connected to a LCMS-8060 (Shimadzu Scientific Instruments, Columbia, MD, USA) with a dual ion source. Chromatographic separation was achieved on an Hypercarb column, 2.1 $\times$ 100 mm, 5 μm (Thermo Fisher Scientifics, Waltham, MA, USA) fluxing mobile phase at a flow rate of 0.3 mL/min under gradient conditions. The mass spectrometer worked with an electrospray ionization source operating both in the negative and positive Multiple Reaction Monitoring mode, quantifying target ions m/z 264$\rightarrow$59.25 for gemcitabine (positive), m/z 502$\rightarrow$158.9 for dFdCTP (negative), m/z 263.1$\rightarrow$220 for dFdU (negative), and m/z 228.1$\rightarrow$112.05 for IS (positive). The lower limit of quantitation (LOQ) was 0.2 ng/mg for gemcitabine, 0.4 ng/mg for dFdU, and 0.5 ng/mg for dFdCTP in tumor samples, while it was 0.01 μg/mL for gemcitabine and 0.02 μg/mL for dFdU in plasma samples.

2.7. Histological and Immunohistochemical Analysis

Tumors were collected, fixed in 10% phosphate-buffered formalin, embedded in paraffin, and cut into 4 μm-thick sections. Sections were stained with Hematoxylin and Eosin (H&E) and Sirius red as described [18]. For immunohistochemical analysis of the tumor vasculature, anti-mouse CD31 antibody SZ31 (Dianova GmbH, Hamburg, Germany) followed by biotin-conjugated goat anti-rat IgG antibody (Vector Laboratories, Burlingame, CA, USA) and streptavidin-alkaline phosphatase conjugate were used [18]. CCN2 expression was evaluated using anti-CCN2 (ab6992, Abcam, Milano, Italy), followed by MACH 4 Universal HRP-Polymer (Biocare Medical, Pacheco, CA, USA) [18]. Negative controls included no-primary antibody control and the use of a specific inhibitory mouse CTGF peptide (ab7861, Abcam). Images (bright field for H&E and CD31 and polarized light for Sirius red) were acquired with Axio Imager Z2 (Zeiss, Felbach, Switzerland). Presence of fibrosis and vascular structures was analyzed using ImageJ software (https://imagej.nih.gov/) and expressed as the percentage of total tumor area. The amount of necrotic tissues in tumors was quantified by blind scoring, exploiting the difference in staining intensity between vital and necrotic tissue.

2.8. Statistical Analysis

Differences in proliferation and tumor growth were analyzed by one-way ANOVA followed by Tukey's or Dunnett's multiple comparisons test. The *p* value < 0.05 was considered significant. Statistical analysis was performed using GraphPad Prism version 8 Software (GraphPad, LaJolla, CA, USA).

3. Results

3.1. Effect of CCN-Targeting Peptides (BLR100 and BLR200) on the Orthotopic Growth of PDAC

The antitumor activity of BLR100 and BLR200 was first investigated in the murine FC1199 cells established from pancreatic tumors in KPC mice. We began by verifying that the tumor model expressed CCN2, the primary target of the peptides. Immunohistochemical analysis of FC1199 tumors grown in the pancreas of syngeneic C57BL/6 mice and characterized by the presence of significant amount of stroma, confirmed that CCN2 was expressed in vivo, both in tumor cells and stroma cells, but was barely detectable in healthy pancreas (Figure 1A–C). In agreement, FC1199 tumor cells in vitro expressed the CCN2 protein: the protein was detectable in the cell lysate and was released in relevant amounts in the conditioned media (Figure 1D). This indicated that FC1199 cells maintain the characteristics of the original KPC tumors, previously reported to produce CCN2 [18,20], and therefore represent a good model to test the therapeutic potential of CCN2-targeting agents, such as BLR100 and BLR200.

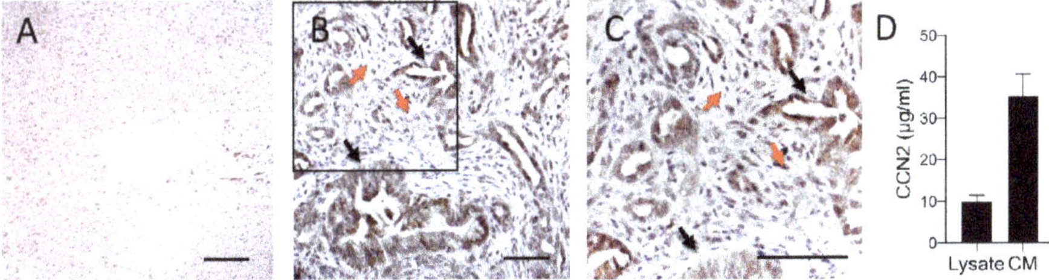

Figure 1. CCN2 is expressed by FC1199 tumors. CCN2 was analyzed by immunohistochemistry (IHC) in healthy murine pancreas (**A**) and FC1199 orthotopic tumors (**B,C**). Black arrows: tubular structures of an adenocarcinoma with cellular atypia and differences in the thickness of the tubular wall. Red arrows: stromal areas. C is a higher magnification of the boxed area in B. CCN2 was expressed by both tumor cells (black arrows) and stroma cells (red arrow), scale bars: 50 μm. (**D**) CCN2 production by FC1199 cells in vitro. CCN2 in the cell lysate and conditioned media was measured by ELISA and expressed as μg/mL.

The activity of BLR100 and BLR200 was then tested on orthotopic, early stage FC1199 tumors, starting treatments 4 days after tumor cell injection, when tumors were histologically detectable (Figure 2A,B). The peptides were initially administered at the dose of 4.5 μg/kg to simulate a "physiological" concentration of CCN3, calculated based on the molar quantity of circulating CCN3 present in a normal mouse [23] and adjusted upward slightly to anticipate increased clearance by the kidneys of a small peptide as compared to full-length CCN3 protein. Preliminary pharmacokinetic study in healthy mice indicated that, following i.p. administration, the peptides distribute to the pancreas, where they were detected in biologically relevant concentrations (not shown). Also, our previous studies in diabetic kidney disease had shown that dosing CCN3 at 3 times per week was sufficient to block and even reverse fibrosis, so a similar schedule was selected [23]. Under these conditions, BLR100 and BLR200 were found to have a moderate, though not statistically significant, antineoplastic activity detectable as early as 10 days from the beginning of treatments (day 14) and becoming more evident at the end of the experiment (day 25, Figure 2C). Notably, the formation of ascites, a typical marker of disease progression, was prevented by both compounds. No mice treated with BLR200 had ascites in the peritoneal cavity while BLR100 treatment clearly reduced the percentage of mice with ascitic fluid (Figure 2D). Although the differences in tumors treated with BLR100 or BLR200 compared to controls did not reach statistical significance, the inhibition induced by the two peptides was reproducible and obtained with multiple preparations of the peptides. To rule out a non-specific activity of the peptides, in a subsequent experiment we used as a control a scramble peptide, consisting of the same amino acids contained in BLR200, but with a random order. Moreover, in this experiment, the dose of the peptides was increased to 10 μg/kg. Tumors treated with BLR100 and BLR200 were significantly smaller than the ones treated with the control peptide, confirming the specificity of the effect (Figure 2E).

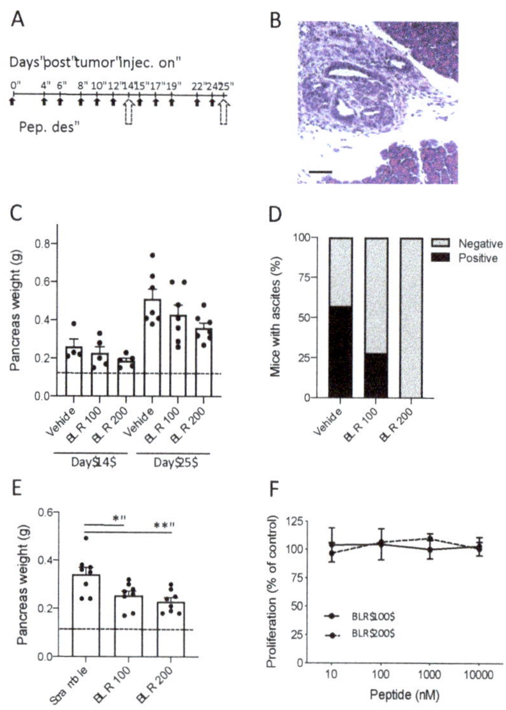

Figure 2. Antineoplastic activity of BLR100 or BLR200 on PDAC. (**A**) Schedule of treatment. The peptides were administered three times per week at 4.5 μg/kg, at the indicated day (black arrows). Dotted arrows indicate mice sacrifice and analysis on days 14 and 25. (**B**) Tumor in the pancreas of mice 4 days after tumor transplantation (H&E staining, scale bar 100 μm). (**C**) Tumor burden (evaluated as pancreas weight) in mice bearing FC1199 orthotopic tumors treated with vehicle, BLR100, or BLR 200 on day 14 (n = 5) or 25 (n = 7). Dotted line indicates the weight of pancreas in healthy mice. (**D**) Effect of treatments on ascites formation. Data are the percentage of mice presenting (black) or not (gray) ascites on day 25 (n = 7). (**E**) Tumor burden in mice treated with a scramble peptide, BLR100, or BLR200 (10 μg/kg) and sacrificed 25 days after tumor transplantation (n = 8). Dotted line indicates the weight of the pancreas in healthy mice. In C and E, data are mean ± SEM. One way ANOVA and Tukey's, ** $p < 0.005$; * $p < 0.05$. (**F**) Effect of BLR100 or BLR200 on FC1199 cell proliferation in vitro. Data are the percentage of control proliferation, mean ± SEM of values from 2 independent experiments.

When examined in vitro, neither BLR100 nor BLR200 affected FC1199 cell proliferation (Figure 2F), indicating that the in vivo activity was not likely due to a direct antiproliferative activity on the tumor cells and supporting the idea that the effect of two peptides on tumor growth was through a modification of the tumor microenvironment.

3.2. BLR100 and BLR200 Reorganize the Tumor Microenvironment

To investigate the possible activity of the compounds on the tumor stroma, we analyzed fibrosis and neo-vascularization in tumors treated or not with BLR100 and BLR200 (4.5 µg/mL). Alterations in the tumor microenvironment were observed in treated tumors, with some differences between the two compounds. Fibrosis, a major hallmark of CCN2 activity, was reduced by treatment with both compounds. Sirius red staining of tumors (Figure 3A) showed a stromal deposition pattern of collagens (predominantly fibers appearing green in polarized light, mainly representing thin collagen fibers), which was reduced by both the peptides, although the reduction was significantly different only in BLR200 treated tumors ($p < 0.05$, Figure 3A). The fiber organization in the tumor microenvironment and the fact that this area is dominated by activated fibroblasts responsible for matrix reorganization suggests an impact of the peptides on fibroblast activity. BLR200 also relevantly reduced the areas of necrosis in the treated tumors compared with controls ($p < 0.05$, Figure 3B), whereas BLR100 was more effective in reducing tumor vascularization ($p < 0.05$, Figure 3C).

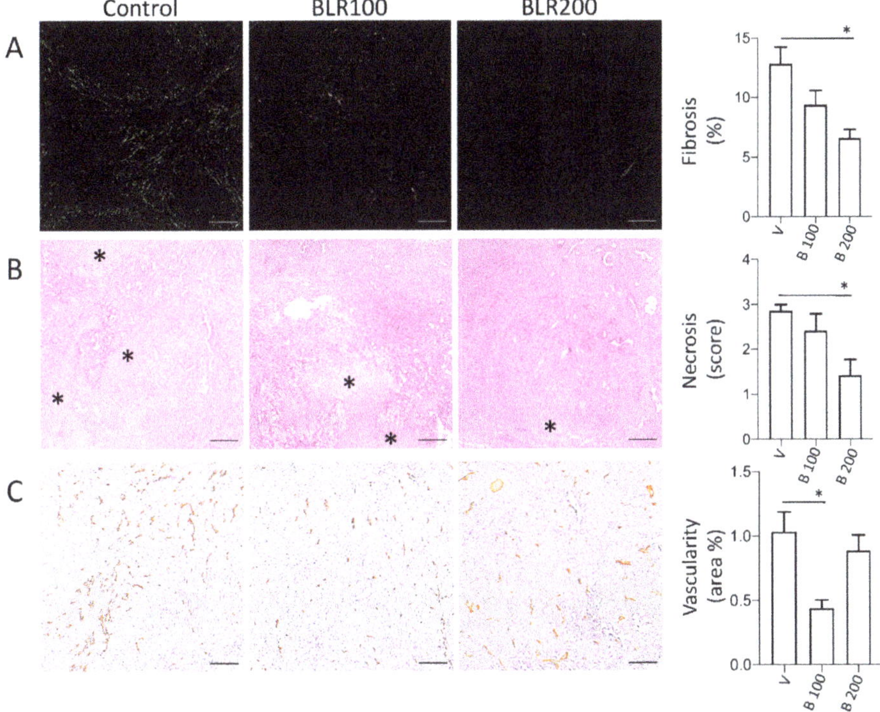

Figure 3. Microenvironmental changes induced by BLR100 or BLR200 on PDAC tumors. Representative images and relative quantification of (**A**) fibrosis (polarized light microscopy analysis of Sirius red stained sections; scale bars, 25 µm), (**B**) necrosis (asterisks, H&E staining; scale bars, 200 µm) and (**C**) tumor vessels (CD31 IHC; scale bars, 200 µm) of control and peptide-treated FC1199 orthotopic tumors. N ≥ 6. One way ANOVA and Dunnett's: * $p < 0.05$.

3.3. Antineoplastic activity of BLR100 and BLR200 in Combination with Chemotherapy

We next investigated the antineoplastic activity of the compounds in combination with the chemotherapeutic agent gemcitabine, standard of care for PDAC. Gemcitabine was administered at the sub-optimal dose of 40 mg/kg and the peptides at the lower dose (4.5 µg/kg). The two peptides were given before and during the administration of the cytotoxic drug, to maximize the effect of microenvironmental changes on drug activity, starting treatments when tumors were palpable (corresponding to a mean pancreas weight of 0.27 g ± 0.05 g as assessed in preliminary experiments) on day 11 (Figure 4A). All treatments were well tolerated and no body weight loss was observed (Figure 4B).

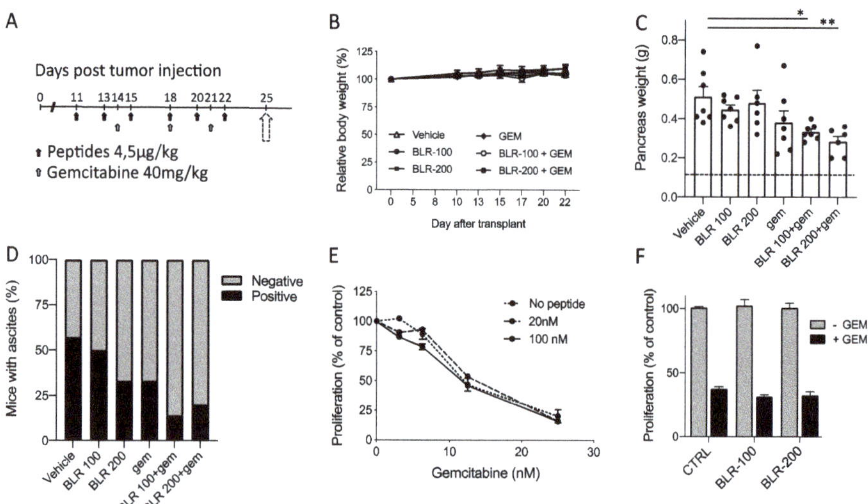

Figure 4. Antineoplastic activity of BLR100 or BLR200 in combination with gemcitabine on PDAC. (**A**) Schedule of treatment. Mice received the peptides (black arrows) and gemcitabine (white arrows) at the indicated times. Dotted arrow indicates mice sacrifice and tumor analysis. (**B**) Effect of the indicated treatments on mouse body weight. Data are expressed as relative body weight, the percentage of mouse weight at the beginning of treatment, mean ± SEM. (**C**) Tumor burden (pancreas weight) in mice bearing FC1199 orthotopic tumors treated with vehicles, peptide BLR100, peptide BLR200, gemcitabine (gem), or gemcitabine in combination with the peptides. Dotted line: weight of pancreas in healthy mice (n = 7, mean ± SEM, One way ANOVA and Dunnett's * $p < 0.05$, ** $p < 0.005$). (**D**) Effect of treatments on ascites formation. Data are the percentage of mice presenting (black) or not (gray) ascites (n = 7). (**E,F**) BLR100 and BLR200 did not affect FC1199 cell response to gemcitabine in vitro. (**E**) Proliferation of FC1199 cell exposed to increasing concentrations of gemcitabine alone or with the peptide (20 and 100 nM). (**F**) Inhibition of FC1199 cell proliferation by gemcitabine (15 nM) in the presence of BLR100 or BLR200 (100 nM). Data are percentage of control, from one experiment representative of two.

Under these conditions, peptide and gemcitabine, as monotherapies, had only a marginal effect on tumor growth, as expected given the low dose of both drugs (equimolar to physiological concentrations of CCN3 for the peptides and suboptimal for gemcitabine) and the late start of peptide treatment (day 11) compared to the earlier treatment of Figure 2. However, tumors in mice treated with either BLR100 or BLR200 in combination with gemcitabine were significantly smaller than controls ($p < 0.05$ and $p < 0.005$ respectively, Figure 4C). Furthermore, the large variation in responsiveness among animals observed in the gemcitabine alone group was reduced when the combination of drugs was used. The combined treatment was also effective in reducing ascites formation, as fewer mice presented ascites in the abdominal cavity compared with mice treated with the vehicle or with the single agents (Figure 4D).

To investigate a possible direct effect of the combination on tumor cells, the effect of BLR100 at the concentrations of 20 nM or 100 nM combined with increasing doses of gemcitabine was evaluated on FC1199 cell proliferation in vitro. Figure 4E shows no difference in proliferation between cells treated with gemcitabine alone or in combination with the compound (IC50 values, in nM, were 13.5 ± 4.6, 13.2 ± 3.9, and 11.9 ± 3.7 for gemcitabine alone, and with BLR100 20 nM and 100 nM, respectively, mean and SD of two experiments). Similar findings were obtained with BLR200 (Figure 4F) indicating that the peptides do not sensitize tumor cells to gemcitabine.

3.4. Pharmacokinetic Studies

Since tumor vascularization, necrosis, and fibrosis are critical factors influencing drug delivery, we investigated whether the improved antineoplastic activity of the combination might be associated with a higher concentration of gemcitabine in the tumors.

Mice bearing FC1199 tumors were treated every other day with BLR200 (10 µg/kg) from day 11 for a total of 4 treatments. On day 17, mice were treated with gemcitabine (100 mg/kg, i.v., single bolus) and sacrificed 30 min and 1 h later. Since in vivo gemcitabine is rapidly metabolized to the active compound dFdCTP or the inactivate metabolite dFdU, we measured the concentration of both gemcitabine and its two metabolites in tumors and plasma by HPLC–MS. In BLR200-treated mice, a small increase of both native gemcitabine and the dFdU metabolite was observed, while the other metabolite, dFdCTP, was undetectable in both groups (Figure 5A,B). Although the changes induced by the peptide did not reach the level for statistical significance in these experiments, they were consistent with peptide-induced biological effect. These findings indicate that a net increase of gemcitabine concentration is not the main determinant of the improved activity of the combination, although we cannot rule out a possible effect of the peptide on the spatial distribution of gemcitabine in the tumor.

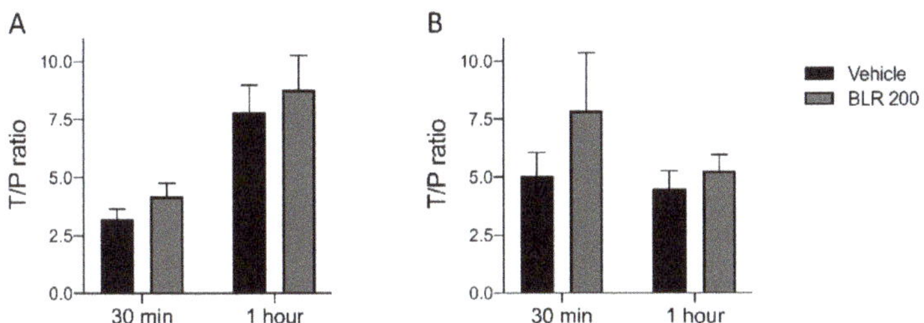

Figure 5. Effect of BLR200 on gemcitabine distribution. Gemcitabine (**A**) and the dFdU metabolite (**B**) were measured by HPLC–MS analysis in plasma and tumors of mice bearing FC1199 orthotopic tumors, treated or not with BLR200. Data are the ratio between tumor (T) and plasma (P) concentration, 30 min and 1 h after gemcitabine administration (mean ± SEM, n = 4).

4. Discussion

The PDAC microenvironment is a recognized determinant of malignant progression and resistance to chemotherapy and has therefore been proposed as a target for treatments. In recent years several agents targeting the tumor microenvironment have therefore been developed, tested in preclinical PDAC models and, in some cases, also in clinical trials in combination with chemotherapy. These studies have had different outcomes depending on the nature of the agent and the clinical setting [28]. A possible reason for the variability of response might be associated with the pathways targeted by those drug candidates and the nature of the PDAC microenvironment, a complex network of interacting molecules that ultimately activate multiple pro- and antitumorigenic signals.

In this study we show that BLR100 and BLR200, two CCN2-targeting, CCN3-derived peptides, can actively modify the PDAC microenvironment and increase tumor response to chemotherapy. The study was conducted in the murine FC1199 tumor model, expressing CCN2 and growing both in vitro and in vivo, hence providing a good tool to study the activity and mechanisms of action of CCN2-targeting compounds. The lack of direct activity of BLR100 and BLR200 on tumor cell proliferation and response to gemcitabine, in vitro, appears to rule out direct cytotoxic and chemosensitizing effects of the peptides on the tumor cells, and points to a microenvironmental-mediated activity. The effect of the compounds on the microenvironment was indeed observed in FC1199 tumors in vivo, where the two peptides reduced tumor angiogenesis, fibrosis, and necrosis. Notably, the two peptides elicited a somewhat different modification of the tumors, with BLR100 appearing to be more effective in inhibiting blood vessel formation, while BLR200 appeared to be more effective at inhibiting fibrosis and necrosis. This different activity might conceivably be due to the origin of the peptides, derived from different domains of CCN3 [24], known to bind to different receptors and to carry out different functions. The finding that both peptides appear effective in PDAC opens two possible scenarios: using each peptide at different stages of disease, or combining the two peptides to simultaneously target multiple aspects of the PDAC microenvironment. Future studies will address these possibilities.

Our results are in line with the role of CCN2 in angiogenesis, vascular remodeling, and fibrosis [19]. These findings show a significant decrease in collagen deposition (predominantly fibers appearing green in polarized light, mainly representing thin collagen fibers secreted by fibroblast/mesenchymal cells), which is highly reduced by the peptides, suggesting an impact of the peptides on fibroblast activity. This is in line with described activity of the peptides that prevent procollagen synthesis by skin fibroblasts (Riser, B.L., U.S. Patent 8518,395, 2013). In preliminary studies, we found no differences in alpha smooth muscle actin (ASMA) positive cells in tumor sections of control and peptide-treated mice (not shown), possibly suggesting that the peptides might act by blocking the CCN2 effect on matrix turnover rather than fibroblast transdifferentiation into myofibroblasts. However, since our study was limited to one marker of activation, and cancer associated fibroblasts are a heterogeneous and plastic population of cells, with different levels of activation, function, and marker expression, further work will be needed to investigate the effect of the peptides on these cells.

It has been shown that CCN2 is essential for vessel maturation [29] and microvascular integrity [30]. In agreement, the CCN2-targeting monoclonal antibody FG-3019 was described as active in reducing the vasculature of the PANC-1 models [19], though not in KPC mice [20]. Two recent studies demonstrated that CCN2 inhibition (FG-3019) or deficiency (CCN2 conditional knockout) resulted in the maintenance of peritoneal function in a chlorhexidine gluconate-induced peritoneal fibrosis model by reducing angiogenesis, fibrosis, and inflammation [31,32]. Along these lines, we observed a substantial inhibition of ascitic fluid accumulation in the peritoneal cavity of mice treated with both the peptides. This finding alone may be of high importance, since in human PDAC the presence of ascites is associated with the late phases of disease progression, and is indicative of poor prognosis [33].

CCN3 is an endogenous inhibitor of CCN2, as it blocks CCN2 functions, including fibrosis, and has therefore been proposed as a model for the design of CCN2 inhibitory agents [24]. Our findings that peptides based on specific amino acid sequences of CCN3 are indeed able to inhibit fibrosis and angiogenesis confirm that CCN3 can antagonize CCN2, even in a tumor setting rich in CCN2, and supports the rationale for the development of such CCN3-based compounds for PDAC therapy. In other cell types, the peptides were able to downregulate CCN2 expression (Riser B.L., U.S. Patent 8518,395, 2013). However, preliminary data indicate that in tumors treated with peptides, CCN2 mRNA (measured by quantitative RT-PCR analysis) and protein (analyzed by IHC) were not decreased compared to controls (data not shown). This, together with the observed ability to reduce fibrosis and angiogenesis (known to be induced by CCN2) indicate that the peptides may act by inhibiting CCN2 activity rather than expression.

Both BLR peptides potentiated the therapeutic effect of gemcitabine. This indicates an additive/synergistic effect of the drugs. The finding that no synergism/additivity was observed in vitro, as the peptides did not increase tumor cell response to gemcitabine, points to a two-compartment activity of the combination, targeting tumor cell proliferation (gemcitabine) and the microenvironment (BLR drugs).

The observed reduction in fibrosis, a major obstacle to drug delivery in tumors [34], prompted us to investigate whether BLR200 might alter gemcitabine distribution. Pharmacokinetic studies indicated a small increase in gemcitabine and dFdU concentration in tumors treated with BLR200. A study using a different CCN2 inhibitory agent, FG-3019, found that the antibody was able to promote KPC tumor response to gemcitabine, without improving drug distribution. However, differently from our findings, CCN2 targeting with FG-3019 did not cause alterations in the tumor microenvironment, suggesting possible differences in the mechanisms of CCN2 inhibition between the antibody and the BLR peptides. In addition to differences in physical properties, FG-3019 would be expected to be highly specific for CCN2, whereas the CCN3 peptides might affect CCN2 and other CCN3-interacting CCNs. Preliminary studies by one of us (BLR) have shown that BLR200 treatment is able to counteract the increase in CCN1 (formerly named Cyr61) in a model of inflammatory skin fibrosis (unpublished data), in agreement with the known role of CCN3 in downregulating CCN1 [35]. Interestingly CCN1 is detected in the early precursor lesions, and intensifies with disease progression in PDAC [36] and has supporting activity in pancreatic cancer growth, invasiveness, and drug resistance [37,38]. Therefore BLR peptides might simultaneously target multiple CCN family members with protumorigenic functions in PDAC.

An additional factor to be considered is the decrease in tumor necrosis following peptide treatment. Necrosis is a major obstacle to drug delivery to tumors. In line with the findings by Cesca et al. [39] we can hypothesize that the presence of large vital areas in BLR200-treated tumors facilitates a more homogeneous distribution of the cytotoxic drug, thus promoting a greater antitumor activity of gemcitabine even in the absence of a marked increase in total drug concentration. Further studies with technologies enabling the spatial analysis of drug distribution in tumors, such as matrix-assisted laser desorption ionization-mass spectrometry imaging, are warranted to address this hypothesis.

In conclusion, this study has shown the therapeutic potential of two CCN3-based peptides in combination with gemcitabine in PDAC. In comparison to other recently developed drugs, the approach proposed here could have a great advantage since CCNs are master regulators of multiple critical disease pathways. Indeed, we observed in this study that BLR100 and BLR200 were able to affect fibrosis deposition, angiogenesis, necrosis, and ascites formation. The different ability of the two peptides to affect distinct aspects of tumor progression suggests that they may be used either in combination or at different stages of the disease. Moreover, the increased activity of the combination regimens of the peptides with gemcitabine opens new perspectives for the use of the peptides to augment the limited efficacy of current chemotherapy and perhaps also immunotherapy in PDAC. Since the increased activity of the combination treatment was obtained with suboptimal doses of gemcitabine, the use of low, "physiological" doses of these peptides might also have interesting clinical implications for the possibility of reducing dosing of the highly toxic gemcitabine while achieving a good therapeutic response. Developing new therapeutic approaches and targeting different features of tumor microenvironment, combined with the standard of care therapy is critical to increase PDAC patient survival.

Author Contributions: A.R. and P.B. performed the experiments and analyzed the data. T.C., A.P., M.Z. performed the pharmacokinetics studies. A.B. performed the IHC analyses. B.L.R. provided the peptide drugs, BLR100 and BLR200, contributed to study design, interpretation of data, and writing of the manuscript. G.T. and D.B. conceived and supervised the project and wrote the paper with input from all authors. All authors have read and agreed to the published version of the manuscript.

Funding: This study was supported by AIRC under IG2019—ID 23443, P.I. D. Belotti and AIRC IG2018–ID 21359, P.I. G. Taraboletti, and partly from BLR Bio. A. Resovi is the recipient of an AIRC fellowship.

Acknowledgments: We thank D.A. Tuveson for kindly providing FC1199 cells.

Conflicts of Interest: B.L.R. is the founder and CEO of BLR Bio and has a commercial interest in the development of the peptides used in this study. The other authors declare no conflict of interest.

References

1. Weniger, M.; Honselmann, K.C.; Liss, A.S. The Extracellular Matrix and Pancreatic Cancer: A Complex Relationship. *Cancers (Basel)* **2018**, *10*, 316. [CrossRef] [PubMed]
2. Merika, E.E.; Syrigos, K.N.; Saif, M.W. Desmoplasia in pancreatic cancer. Can we fight it? *Gastroenterol. Res. Pract.* **2012**, *2012*, 781765. [CrossRef] [PubMed]
3. Lafaro, K.J.; Melstrom, L.G. The Paradoxical Web of Pancreatic Cancer Tumor Microenvironment. *Am. J. Pathol.* **2019**, *189*, 44–57. [CrossRef] [PubMed]
4. Neesse, A.; Algül, H.; Tuveson, D.A.; Gress, T.M. Stromal biology and therapy in pancreatic cancer: A changing paradigm. *Gut* **2015**, *64*, 1476–1484. [CrossRef]
5. Belotti, D.; Foglieni, C.; Resovi, A.; Giavazzi, R.; Taraboletti, G. Targeting angiogenesis with compounds from the extracellular matrix. *Int. J. Biochem. Cell Biol.* **2011**, *43*, 1674–1685. [CrossRef] [PubMed]
6. Jun, J.-I.; Lau, L.F. Taking aim at the extracellular matrix: CCN proteins as emerging therapeutic targets. *Nat. Rev. Drug Discov.* **2011**, *10*, 945–963. [CrossRef]
7. Lau, L.F. Cell surface receptors for CCN proteins. *J. Cell Commun. Signal.* **2016**, *10*, 121–127. [CrossRef]
8. Makino, Y.; Hikita, H.; Kodama, T.; Shigekawa, M.; Yamada, R.; Sakamori, R.; Eguchi, H.; Morii, E.; Yokoi, H.; Mukoyama, M.; et al. CTGF Mediates Tumor-Stroma Interactions between Hepatoma Cells and Hepatic Stellate Cells to Accelerate HCC Progression. *Cancer Res.* **2018**, *78*, 4902–4914. [CrossRef]
9. Banerjee, S.K.; Maity, G.; Haque, I.; Ghosh, A.; Sarkar, S.; Gupta, V.; Campbell, D.R.; Von Hoff, D.; Banerjee, S. Human pancreatic cancer progression: An anarchy among CCN-siblings. *J. Cell Commun. Signal.* **2016**, *10*, 207–216. [CrossRef]
10. Wenger, C.; Ellenrieder, V.; Alber, B.; Lacher, U.; Menke, A.; Hameister, H.; Wilda, M.; Iwamura, T.; Beger, H.G.; Adler, G.; et al. Expression and differential regulation of connective tissue growth factor in pancreatic cancer cells. *Oncogene* **1999**, *18*, 1073–1080. [CrossRef]
11. di Mola, F.F.; Friess, H.; Martignoni, M.E.; Di Sebastiano, P.; Zimmermann, A.; Innocenti, P.; Graber, H.; Gold, L.I.; Korc, M.; Büchler, M.W. Connective tissue growth factor is a regulator for fibrosis in human chronic pancreatitis. *Ann. Surg.* **1999**, *230*, 63–71. [CrossRef] [PubMed]
12. Riser, B.L.; Denichilo, M.; Cortes, P.; Baker, C.; Grondin, J.M.; Yee, J.; Narins, R.G. Regulation of connective tissue growth factor activity in cultured rat mesangial cells and its expression in experimental diabetic glomerulosclerosis. *J. Am. Soc. Nephrol.* **2000**, *11*, 25–38. [PubMed]
13. Rayego-Mateos, S.; Morgado-Pascual, J.L.; Rodrigues-Diez, R.R.; Rodrigues-Diez, R.; Falke, L.L.; Mezzano, S.; Ortiz, A.; Egido, J.; Goldschmeding, R.; Ruiz-Ortega, M. Connective tissue growth factor induces renal fibrosis via epidermal growth factor receptor activation. *J. Pathol.* **2018**, *244*, 227–241. [CrossRef] [PubMed]
14. Wang, X.; Cui, H.; Wu, S. CTGF: A potential therapeutic target for Bronchopulmonary dysplasia. *Eur. J. Pharm.* **2019**, *860*, 172588. [CrossRef] [PubMed]
15. Dorn, L.E.; Petrosino, J.M.; Wright, P.; Accornero, F. CTGF/CCN2 is an autocrine regulator of cardiac fibrosis. *J. Mol. Cell. Cardiol.* **2018**, *121*, 205–211. [CrossRef]
16. Kubota, S.; Takigawa, M. Cellular and molecular actions of CCN2/CTGF and its role under physiological and pathological conditions. *Clin. Sci.* **2015**, *128*, 181–196. [CrossRef]
17. Bennewith, K.L.; Huang, X.; Ham, C.M.; Graves, E.E.; Erler, J.T.; Kambham, N.; Feazell, J.; Yang, G.P.; Koong, A.; Giaccia, A.J. The role of tumor cell-derived connective tissue growth factor (CTGF/CCN2) in pancreatic tumor growth. *Cancer Res.* **2009**, *69*, 775–784. [CrossRef]
18. Resovi, A.; Bani, M.R.; Porcu, L.; Anastasia, A.; Minoli, L.; Allavena, P.; Cappello, P.; Novelli, F.; Scarpa, A.; Morandi, E.; et al. Soluble stroma-related biomarkers of pancreatic cancer. *Embo Mol. Med.* **2018**, *10*, e8741. [CrossRef]
19. Aikawa, T.; Gunn, J.; Spong, S.M.; Klaus, S.J.; Korc, M. Connective tissue growth factor-specific antibody attenuates tumor growth, metastasis, and angiogenesis in an orthotopic mouse model of pancreatic cancer. *Mol. Cancer* **2006**, *5*, 1108–1116. [CrossRef]

20. Neesse, A.; Frese, K.K.; Bapiro, T.E.; Nakagawa, T.; Sternlicht, M.D.; Seeley, T.W.; Pilarsky, C.; Jodrell, D.I.; Spong, S.M.; Tuveson, D.A. CTGF antagonism with mAb FG-3019 enhances chemotherapy response without increasing drug delivery in murine ductal pancreas cancer. *Proc. Natl. Acad. Sci. USA* **2013**, *110*, 12325–12330. [CrossRef] [PubMed]
21. Riser, B.L.; Najmabadi, F.; Perbal, B.; Rambow, J.A.; Riser, M.L.; Sukowski, E.; Yeger, H.; Riser, S.C.; Peterson, D.R. CCN3/CCN2 regulation and the fibrosis of diabetic renal disease. *J. Cell Commun. Signal.* **2010**, *4*, 39–50. [CrossRef] [PubMed]
22. Riser, B.L.; Najmabadi, F.; Perbal, B.; Peterson, D.R.; Rambow, J.A.; Riser, M.L.; Sukowski, E.; Yeger, H.; Riser, S.C. CCN3 (NOV) is a negative regulator of CCN2 (CTGF) and a novel endogenous inhibitor of the fibrotic pathway in an in vitro model of renal disease. *Am. J. Pathol.* **2009**, *174*, 1725–1734. [CrossRef] [PubMed]
23. Riser, B.L.; Najmabadi, F.; Garchow, K.; Barnes, J.L.; Peterson, D.R.; Sukowski, E.J. Treatment with the matricellular protein CCN3 blocks and/or reverses fibrosis development in obesity with diabetic nephropathy. *Am. J. Pathol.* **2014**, *184*, 2908–2921. [CrossRef] [PubMed]
24. Riser, B.L.; Barnes, J.L.; Varani, J. Balanced regulation of the CCN family of matricellular proteins: A novel approach to the prevention and treatment of fibrosis and cancer. *J. Cell Commun. Signal.* **2015**, *9*, 327–339. [CrossRef] [PubMed]
25. Hingorani, S.R.; Wang, L.; Multani, A.S.; Combs, C.; Deramaudt, T.B.; Hruban, R.H.; Rustgi, A.K.; Chang, S.; Tuveson, D.A. Trp53R172H and KrasG12D cooperate to promote chromosomal instability and widely metastatic pancreatic ductal adenocarcinoma in mice. *Cancer Cell* **2005**, *7*, 469–483. [CrossRef]
26. Workman, P.; Aboagye, E.O.; Balkwill, F.; Balmain, A.; Bruder, G.; Chaplin, D.J.; Double, J.A.; Everitt, J.; Farningham, D.a.H.; Glennie, M.J.; et al. Guidelines for the welfare and use of animals in cancer research. *Br. J. Cancer* **2010**, *102*, 1555–1577. [CrossRef]
27. Bapiro, T.E.; Richards, F.M.; Goldgraben, M.A.; Olive, K.P.; Madhu, B.; Frese, K.K.; Cook, N.; Jacobetz, M.A.; Smith, D.-M.; Tuveson, D.A.; et al. A novel method for quantification of gemcitabine and its metabolites 2′,2′-difluorodeoxyuridine and gemcitabine triphosphate in tumour tissue by LC-MS/MS: Comparison with (19)F NMR spectroscopy. *Cancer Chemother. Pharm.* **2011**, *68*, 1243–1253. [CrossRef]
28. Liang, C.; Shi, S.; Meng, Q.; Liang, D.; Ji, S.; Zhang, B.; Qin, Y.; Xu, J.; Ni, Q.; Yu, X. Complex roles of the stroma in the intrinsic resistance to gemcitabine in pancreatic cancer: Where we are and where we are going. *Exp. Mol. Med.* **2017**, *49*, e406. [CrossRef]
29. Hall-Glenn, F.; De Young, R.A.; Huang, B.-L.; van Handel, B.; Hofmann, J.J.; Chen, T.T.; Choi, A.; Ong, J.R.; Benya, P.D.; Mikkola, H.; et al. CCN2/connective tissue growth factor is essential for pericyte adhesion and endothelial basement membrane formation during angiogenesis. *PLoS ONE* **2012**, *7*, e30562. [CrossRef]
30. Chaqour, B. Caught between a "Rho" and a hard place: Are CCN1/CYR61 and CCN2/CTGF the arbiters of microvascular stiffness? *J. Cell Commun. Signal.* **2019**. [CrossRef]
31. Toda, N.; Mori, K.; Kasahara, M.; Koga, K.; Ishii, A.; Mori, K.P.; Osaki, K.; Mukoyama, M.; Yanagita, M.; Yokoi, H. Deletion of connective tissue growth factor ameliorates peritoneal fibrosis by inhibiting angiogenesis and inflammation. *Nephrol. Dial. Transpl.* **2018**, *33*, 943–953. [CrossRef]
32. Sakai, N.; Nakamura, M.; Lipson, K.E.; Miyake, T.; Kamikawa, Y.; Sagara, A.; Shinozaki, Y.; Kitajima, S.; Toyama, T.; Hara, A.; et al. Inhibition of CTGF ameliorates peritoneal fibrosis through suppression of fibroblast and myofibroblast accumulation and angiogenesis. *Sci. Rep.* **2017**, *7*, 5392. [CrossRef] [PubMed]
33. Zervos, E.E.; Osborne, D.; Boe, B.A.; Luzardo, G.; Goldin, S.B.; Rosemurgy, A.S. Prognostic significance of new onset ascites in patients with pancreatic cancer. *World J. Surg. Oncol.* **2006**, *4*, 16. [CrossRef] [PubMed]
34. Olive, K.P.; Jacobetz, M.A.; Davidson, C.J.; Gopinathan, A.; McIntyre, D.; Honess, D.; Madhu, B.; Goldgraben, M.A.; Caldwell, M.E.; Allard, D.; et al. Inhibition of Hedgehog signaling enhances delivery of chemotherapy in a mouse model of pancreatic cancer. *Science* **2009**, *324*, 1457–1461. [CrossRef] [PubMed]
35. Liu, J.; Ren, Y.; Kang, L.; Zhang, L. Overexpression of CCN3 inhibits inflammation and progression of atherosclerosis in apolipoprotein E-deficient mice. *PLoS ONE* **2014**, *9*, e94912. [CrossRef]
36. Haque, I.; Mehta, S.; Majumder, M.; Dhar, K.; De, A.; McGregor, D.; Van Veldhuizen, P.J.; Banerjee, S.K.; Banerjee, S. Cyr61/CCN1 signaling is critical for epithelial-mesenchymal transition and stemness and promotes pancreatic carcinogenesis. *Mol. Cancer* **2011**, *10*, 8. [CrossRef]

37. Maity, G.; Ghosh, A.; Gupta, V.; Haque, I.; Sarkar, S.; Das, A.; Dhar, K.; Bhavanasi, S.; Gunewardena, S.S.; Von Hoff, D.D.; et al. CYR61/CCN1 Regulates dCK and CTGF and Causes Gemcitabine-resistant Phenotype in Pancreatic Ductal Adenocarcinoma. *Mol. Cancer* **2019**, *18*, 788–800. [CrossRef]
38. Wang, X.; Deng, Y.; Mao, Z.; Ma, X.; Fan, X.; Cui, L.; Qu, J.; Xie, D.; Zhang, J. CCN1 promotes tumorigenicity through Rac1/Akt/NF-κB signaling pathway in pancreatic cancer. *Tumour. Biol.* **2012**, *33*, 1745–1758. [CrossRef]
39. Cesca, M.; Morosi, L.; Berndt, A.; Fuso Nerini, I.; Frapolli, R.; Richter, P.; Decio, A.; Dirsch, O.; Micotti, E.; Giordano, S.; et al. Bevacizumab-Induced Inhibition of Angiogenesis Promotes a More Homogeneous Intratumoral Distribution of Paclitaxel, Improving the Antitumor Response. *Mol. Cancer* **2016**, *15*, 125–135. [CrossRef]

Article

Evaluation of Glycosylated PTGS2 in Colorectal Cancer for NSAIDS-Based Adjuvant Therapy

Roberta Venè [1], Delfina Costa [1], Raffaella Augugliaro [2], Sebastiano Carlone [3], Stefano Scabini [4], Gianmaria Casoni Pattacini [4], Maurizio Boggio [5], Simonetta Zupo [6], Federica Grillo [5,7], Luca Mastracci [5,7], Francesca Pitto [5], Simona Minghelli [8], Nicoletta Ferrari [1], Francesca Tosetti [1], Emanuele Romairone [9], Maria C. Mingari [2,10], Alessandro Poggi [1,†] and Roberto Benelli [2,*,†]

[1]   OU Molecular Oncology & Angiogenesis, IRCCS Ospedale Policlinico San Martino, Largo Rosanna Benzi 10, 16132 Genoa, Italy; roberta.vene@hsanmartino.it (R.V.); delfina.costa@hsanmartino.it (D.C.); nicoletta.ferrari@hsanmartino.it (N.F.); francesca.tosetti@hsanmartino.it (F.T.); alessandro.poggi@hsanmartino.it (A.P.)

[2]   OU Immunology, IRCCS Ospedale Policlinico San Martino, Largo Rosanna Benzi 10, 16132 Genoa, Italy; raffaella.augugliaro@hsanmartino.it (R.A.); mariacristina.mingari@unige.it (M.C.M.)

[3]   OU Cell Biology, IRCCS Ospedale Policlinico San Martino, Largo Rosanna Benzi 10, 16132 Genoa, Italy; sebastiano.carlone@hsanmartino.it

[4]   OU Oncologic Surgery and Implantable Systems, IRCCS Ospedale Policlinico San Martino, Largo Rosanna Benzi 10, 16132 Genoa, Italy; stefanoscabini@libero.it (S.S.); gianmaria.casonipattacini@gmail.com (G.C.P.)

[5]   OU Pathology, IRCCS Ospedale Policlinico San Martino, Largo Rosanna Benzi 10, 16132 Genoa, Italy; maurizio.boggio@hsanmartino.it (M.B.); federica.grillo@unige.it (F.G.); mastracc@hotmail.com (L.M.); pitto.francesca@yahoo.it (F.P.)

[6]   OU Molecular Diagnostics, IRCCS Ospedale Policlinico San Martino, Largo Rosanna Benzi 10, 16132 Genoa, Italy; simonetta.zupo@hsanmartino.it

[7]   Department of Surgical Science and Integrated Diagnostics, University of Genoa, 16132 Genoa, Italy

[8]   Clinical and Experimental Immunology lab, Ospedale G. Gaslini, 16147 Genoa, Italy; simoming@tin.it

[9]   Department of General Surgery, Asl3, Ospedale Villa Scassi, 16149 Genoa, Italy; emanuele.romairone@asl3.liguria.it

[10] Department of Experimental Medicine (DIMES), University of Genoa, 16132 Genoa, Italy

*   Correspondence: roberto.benelli@hsanmartino.it; Tel.: +39-010-555-8234

†   These authors have contributed equally.

Received: 28 November 2019; Accepted: 6 March 2020; Published: 11 March 2020

Abstract: Observational/retrospective studies indicate that prostaglandin-endoperoxide synthase-2 (PTGS2) inhibitors could positively affect colorectal cancer (CRC) patients' survival after diagnosis. To obtain an acceptable cost/benefit balance, the inclusion of PTGS2 inhibitors in the adjuvant setting needs a selective criterion. We quantified the 72 kDa, CRC-associated, glycosylated form of PTGS2 in 100 frozen CRC specimens and evaluated PTGS2 localization by IHC in the same tumors, scoring tumor epithelial-derived and stroma-derived fractions. We also investigated the involvement of interleukin-1 beta (IL1β) in PTGS2 induction, both in vitro and in CRC lysates. Finally, we used overall survival (OS) as a criterion for patient selection. Glycosylated PTGS2 can be quantified with high sensibility in tissue lysates, but the expression in both tumor and stromal cells limits its use for predictive purposes. Immunohistochemistry (IHC) analysis indicates that stromal PTGS2 expression could exert a protective role on patient OS. Stromal PTGS2 was prevalently expressed by cancer-associated fibroblasts exerting a barrier function near the gut lumen, and it apparently favored the antitumor M1 macrophage population. IL1β was directly linked to gPTGS2 expression both in vitro and in tumors, but its activity was apparently prevalent on the stromal cell population. We suggest that stromal PTGS2 could exert a positive effect on patients OS when expressed in the luminal area of the tumor.

Keywords: prostaglandin-endoperoxide synthase-2/Cyclooxygenase 2 (PTGS2/COX-2); colorectal cancer (CRC); non-steroidal anti-inflammatory drugs (NSAIDS); cancer-associated fibroblasts (CAF); interleukin-1 beta (IL1β); macrophages

1. Introduction

Prostaglandin-endoperoxide synthase-2 (PTGS2), one of the key enzymes mediating prostaglandins neosynthesis, is typically induced by inflammatory stimuli and expressed by tumor epithelial cells in about 74–78% of colorectal cancer (CRC) (see [1] for review). PTGS2 exists both as a rapidly-degraded 68 kDa unglycosylated form, with increased catalytic activity, and a more stable, endoplasmic reticulum-associated, 72 kDa glycosylated form (gPTGS2) [2]. While unglycosylated PTGS2 can be detected in the normal mucosa, gPTGS2 is typically associated with CRC.

PTGS2 has been considered an ideal target for colorectal tumor chemoprevention [3,4], but the cardiotoxicity associated to the specific PTGS2 inhibitor Celecoxib determined an unfavorable cost/benefit ratio for the chemoprevention of the normal population. On the contrary, the inhibition of PTGS2 in the adjuvant setting could be beneficial for CRC patients. Three independent observational studies by Ng, Hua, and Friis indicated an increased survival for long-term, regular users of non-steroidal anti-inflammatory drugs (NSAIDS) after CRC diagnosis [5–7], of which specific PTGS2 inhibitors were found to be the most active. At present, Celecoxib is prospectively tested in patients with resected stage III colon cancer and treated with adjuvant FOLFOX chemotherapy (https: //clinicaltrials.gov/ct2/show/NCT01150045). This phase III, multicenter trial will give a fundamental hint for the rational use of PTGS2 inhibitors in advanced CRC. Nevertheless, a limit of this study and of future applications is the lack of criteria for patient selection. The influence of tumor PTGS2 expression on CRC patient prognosis is difficult to interpret [8–11]. Moreover, despite the influence of tumor stroma and leucocyte infiltration in CRC progression [12–14], previous studies did not evaluate the influence of PTGS2 expressed by non-tumor cells on patient prognosis. We here quantified the 72 kDa gPTGS2 in 100 primary CRC lysates as a proof of principle for the identification of patients that could benefit from NSAIDS treatment after surgery. PTGS2 levels were also evaluated by the same antibody in immunohistochemistry (IHC), distinguishing tumor-derived from stroma-derived PTGS2. We also evaluated IL1β as a candidate of inflammation-driven stromal PTGS2 expression. PTGS2 was finally correlated with patient prognosis to evaluate its association with CRC aggressiveness.

2. Materials and Methods

2.1. Patients

The study was conducted in accordance with the Declaration of Helsinki, and the protocol was approved by the Ethics Committee of San Martino Hospital (Ethical code number: n°4/2011). All subjects were recruited at the unit of Oncologic Surgery and Implantable Systems after giving their informed consent. A total of 100 patients subjected to surgical resection of CRC by dedicated surgeons were included (49 males and 51 females; median age 70 years). All patients underwent surgery as the first curative treatment. Tumors were located in the ascending (41), transverse (2), descending (17), sigmoid colon (19), and rectum (21), and they were staged as I (14), II (34), III (39), and IV (13) according to Union for International Cancer Control (UICC) 2009 classification.

2.2. Specimens Collection and Processing

Each surgical specimen was collected within 20 min after resection and evaluated by an expert pathologist who collected a representative fragment of the invasive tumor and a strip of normal mucosa, which was sampled at least 10 cm from the tumor mass. Each sample was placed in Safe-Lock tubes (Eppendorf Srl, Milan, Italy) with 80 µL of RIPA buffer containing sodium orthovanadate

(OV) 1 mM, dithiothreitol (DTT) 1 mM and a protease inhibitor cocktail 1:100 (Sigma-Aldrich Italia, Milan, Italy, P8340), and stored at −80 °C. Frozen tissues were thawed on ice and minced with sharp scissors, adding 100 µL of fresh RIPA buffer (with OV, DTT and protease inhibitors) to each sample. After 90 min incubation on ice, samples were potterized and centrifuged (24,000× *g*, 4 °C). Supernatants were collected and protein content was quantified by the DC protein assay (Bio-Rad Laboratories Srl, Milan, Italy).

2.3. Cell Lines

CaCo2 and HT29 (PTGS2 positive), SW480 and HCT15 (PTGS2 negative), DLD1, SW620, LS180 human CRC cell lines (obtained from the Biological Bank of our institute, http://www.iclc.it), and MF2T colon fibroblast primary cell culture [15] were cultured in RPMI 10% FCS.

2.4. Western Blot

Additional information about antibody selection and Western blot quantification of PTGS2 is available in Supplementary method S1.

All 100 samples from normal mucosa and cancer tissue (= 200 samples) were analyzed. Total proteins (30 µg/lane) were resolved on 10% SDS PAGE precast gels (Thermo Scientific) and blotted on PVDF membranes (GE-healthcare Italia, Milan, Italy). Anti PTGS2 (D5H5) rabbit mAb and HRP-conjugated (goat anti-rabbit 7074S) secondary antibody were from Cell Signaling Technology, Leiden, The Netherlands). HRP-conjugated anti beta-actin (13E5 rabbit mAb, Cell Signaling Technology) was used as loading control. Protein bands were detected by a chemiluminescent HRP substrate (Immobilon Western, Merk Life Science Srl, Milan, Italy) and acquired by a C-Digit blot scanner (LI-COR, Bad Homburg, Germany). Only the 72 kDa gPTGS2 band was quantified by the Image Studio 4.0 software. All blots were normalized against two CaCo2 internal standards (10 and 30 µg) loaded in each blot. gPTGS2 relative values were normalized extracting the cubic root (CBRT) of each value to allow the application of parametric statistics. Human PTGS2 standard (cat. 100200-4 Alpha Diagnostic International, San Antonio, TX, USA) was used to estimate gPTGS2 concentration in 30 µg of total tissue lysate.

In vitro studies: in the first test, MF2T primary fibroblasts from human colon were serum-starved for 48 hours and treated with interleukin 8 IL8/CXCL8 (Peprotech, London, UK 10 ng/mL), PGE2 (Cayman, Ann Arbor, MI, USA, 100 nM), growth-regulated oncogene beta GROβ/CXCL2 (Peprotech, 10 ng/mL), interleukin-1 beta (IL1β, Peprotech, 0.1 ng/mL), or epithelial growth factor EGF (Peprotech, 10 ng/mL) for 24 hours to verify the leading role of IL1β in PTGS2 induction (tested in duplicate). In the second test, CRC cell lines were treated with 0.1 ng/mL IL1β in the same conditions of MF2T (tested in duplicate). Cells were scraped, washed once in PBS, and immediately lysed in RIPA buffer. Western blot was run as reported above for tissue samples.

2.5. Immunohistochemistry

A pathologist identified the representative paraffin-embedded tumor sample for each of the 100 cases analyzed by western blot (WB). Four µm thick sections were cut and mounted on Superfrost slides (Thermo-Fisher Scientific Italia, Milan, Italy). IHC was carried out with the automated BenchMark Ultra Immunostainer®(Ventana Medical Systems Tucson, Arizona, USA). Primary, anti-PTGS2 (D5H5) rabbit mAb was used at 1:100 final dilution and detected by Ultraview universal DAB detection kit (Ventana Medical Systems).

PTGS2 was scored by a trained pathologist as the percent of positive cells for both stromal and tumor epithelial cells. Necrotic tissue was excluded from evaluation. Percentages were subdivided into three categories (Low, Medium, High) for Kaplan–Meier analysis. Positivity was classified as follows: negative or barely distinguishable staining, or ≤5% = Low; >5% to ≤20% = Medium; >20% = High.

PTGS2 positive cells in tumor stroma were further characterized either as serial sections stained with DAB for bright field microscopy (CD68 and CD163 positive macrophages: 85 hot-spots in 33 samples), or by the double-fluorescent staining of single sections (Vimentin-positive mesenchymal cells, 14 samples). Samples were probed with the following antibodies: anti PTGS2 D5H5 mAb, anti vimentin (1:100, Thermo-Fisher Scientific Italia), anti CD68 (1:500, BioCare, Pacheco, CA, USA), anti CD163 (1:300, BioCare), using a Leica Bond-RX immunostainer. Images were captured by a Leica AT2 scanner (bright field) or Leica DM-LB2 microscope (Leica Biosystems, Milancity, Italy) equipped with a GXCam-U3-18 camera (fluorescence). The percent of PTGS2, CD68, and CD163-positive cells in hot spots was quantified by the Image Scope 12.3 software (Leica). PTGS2-vimentin fluorescent co-localization was analyzed by the JACoP plug-in of ImageJ (https://imagej.nih.gov/ij/plugins/track/jacop2.html).

Multiplexed IHC on single tissue sections was performed using AEC (Enzo life sciences, Farmingdale, NY, USA) as the chromogenic substrate. Anti mannose receptor 1 (MRC1) (Sigma-Aldrich Italia, prestige antibody AMAb90746; 1:5000 dilution), anti inducible nitric oxide synthase (iNOS) (Thermo-Fisher Scientific Italia, PA3-030A; 1:600 dilution) and anti arginase 1 (ARG1) (Sigma-Aldrich Italia, prestige antibody HPA003595; 1:2000 dilution) were tested in addition to anti CD68 and anti CD163. The first multiplex tested CD68–iNOS–PTGS2 consecutively (7 samples–36 fields–108 images), the second multiplex ARG1–MRC1–CD163–PTGS2 (7 samples–44 fields–176 images). Controls of complete destaining were performed using only the secondary antibody between iNOS and PTGS2 in the first series and MRC1 and CD163 in the second series: no signal was detected. After AT2 slide scanning of each single staining, slides were destained by ETOH washings (5 min 50% ETOH, 10 min 100% ETOH, 5 min 50% ETOH) and antibodies were removed by a guanidine-based stripping solution (6 M Gn-HCl, 0.2% NP-40, 10 mM DTT, 20 mM Tris-HCl, pH 7.5; 37 °C, 20 min). After extensive washing in running tap water, (5′) slides were probed with the successive primary antibody. Hematoxylin was used for nuclear counterstaining only in the first run. AEC positive staining was identified in each slide using Image-J color thresholds and extracted as 8-bit black&white mask (see Supplementary Figure S1), the colocalization of markers (1:1) was quantified by JACoP plug in.

2.6. IL1β ELISA

IL1β was quantified in tumor tissue lysates from 60 unselected cases by RayBio Human IL1 beta ELISA (RayBiotech, Peachtree Corners, GA, USA, ELH-IL1β) according to the manufacturer's instructions; 15 µg of tissue lysate in 100 µL of diluent was plated in each well (test run in duplicate).

2.7. Statistics

All analyses were performed using the free statistical software EZR 1.41 (http://www.jichi.ac.jp/saitama-sct/SaitamaHP.files/statmed.html). The analysis of variance among three or more groups of data was performed by one-way ANOVA or Kruskal–Wallis test. Correlations were calculated by Pearson's or Spearman's test. Nominal data distribution was analyzed by Fisher's exact test. Patients' survival was analyzed by Kaplan–Meier analysis and a Long-rank test. A $p \leq 0.05$ was considered statistically significant.

3. Results

3.1. gPTGS2 Quantification in 100 CRC Lysates and its Relation to Tissue PTGS2

gPTGS2 was detectable by WB in 96/100 CRC (Figure 1a,c) (median = 156.86 pg, mean = 293.3 pg, range 0.00–1515.64 pg of protein, in 30 µg of tissue lysate, according to the hu PTGS2 standard) and in 11/100 of matched normal mucosa (median = 0.00 pg, mean = 0.003 pg, range 0.00–79.8 pg of protein, in 30 µg of tissue lysate). Compared to other studies (see Supplementary Table S1), this is a high detection rate. The replicate of WB analysis on 60 CRC (Figure 1b) showed a high correlation (Pearson's correlation r = 0.907, p = 0.000000000000000000000217, ensuring sufficient reproducibility.

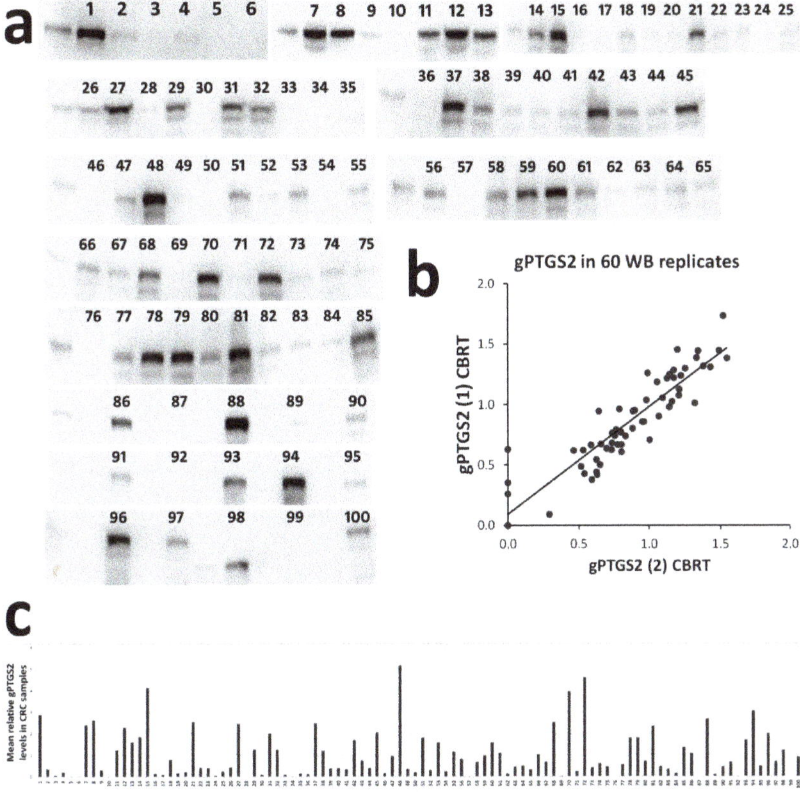

Figure 1. Western blot quantification of glycosylated prostaglandin-endoperoxide synthase-2 (gPTGS2) in colorectal cancer (CRC) lysates: (**a**) c-digit-extracted pseudo-images showing gPTGS2 signal in 100 CRC samples; (**b**) reproducibility of PTGS2 quantification, by replicated WB analysis, in 60 CRC samples; (**c**) relative quantification of gPTGS2 levels in CRC samples.

PTGS2 was also evaluated by IHC in 100 matched CRC paraffin embedded tissues, using the same primary antibody. Tumor-associated and stroma-associated PTGS2 were scored independently. The correlation coefficient of tumor PTGS2 compared with stromal PTGS2 was 0.334 (Spearman's rank, $p < 0.001$). Thus, the contemporary presence of high or low PTGS2 levels in the tumor and stromal populations of the same sample was apparently infrequent in our cohort, suggesting the existence of distinct mechanisms of PTGS2 induction in the different cell populations of the same tumor. In tissue lysates, both tumor and stromal cells contributed to total gPTGS2 levels, showing a directly proportional correspondence with IHC data (Figure 2a).

PTGS2-positive cells of the stromal component almost invariably localized in the luminal area of the tumor, with a strong intensity of staining. These cells frequently lined the limit between living tissue and necrotic areas or surrounded crypts of the outer epithelial border (Figure 2b), suggesting a protective function. In CRCs with medium–high PTGS2 epithelial staining, an irregular distribution of positive areas was observed (Figure 2b).

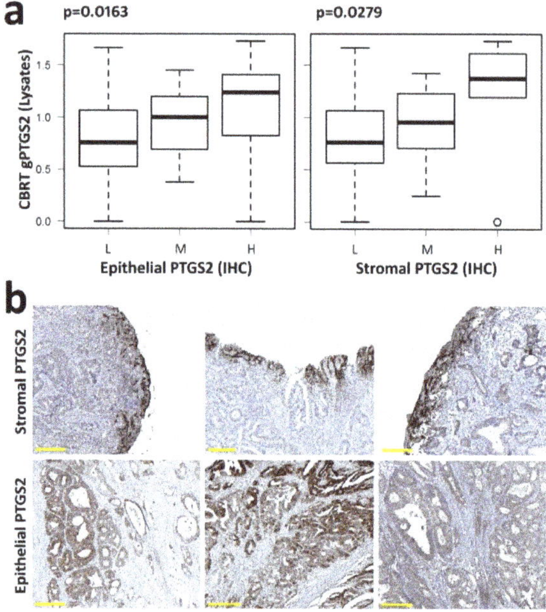

Figure 2. gPTGS2 is expressed in tumor stroma and tumor epithelial cells: (**a**) gPTGS2 levels, quantified in 100 CRC lysates by WB, show a directly proportional correlation with both epithelial and stromal PTGS2 scored by IHC (L = low, M = intermediate, H = high PTGS2 expression); (**b**) PTGS2 positive stromal cells show a bright staining and localize close to the outer mucosal layer of the tumor (upper row), suggesting a protective, barrier function. Tumor epithelial positivity can be observed in all tumor areas, with variable intensity of staining and localization (lower row). Yellow scale bar = 200 μm.

3.2. Identification of gPTGS2 Positive Cells in the Stromal Component

As in our CRC cohort, PTGS2-positive stromal populations with a luminal distribution were previously observed in colon adenomas: Chapple and Bamba independently attributed PTGS2 positivity to macrophages, according to cell morphology or CD68 expression [16,17]. Tumor-infiltrating macrophages have been classified as M1 (antitumor) or M2 (protumor) according to the co-expression of CD68, iNOS or MRC1/CD206, CD163, Arg1, and other markers in in vitro models. In human pathology, this subdivision is an oversimplification, and these markers can be expressed or downregulated in macrophages with high plasticity, according to different microenvironmental stimuli [18]. In Apc (Min/+) mice, the inhibition of PTGS2 reduces the M2 component [19]; thus, the expression of PTGS2 in CRC macrophages could be associated to the induction of a prevalent M2 phenotype. On the other hand, PGE2 is able to induce M1 differentiation in other mouse models [20], suggesting a positive influence of PTGS2 on the M1 component. We first tested the correspondence of PTGS2 expression with CD68 and CD163 on serial sections, assuming that CD68 positivity would indicate the total macrophage population, while CD163 would indicate the M2 component. The comparison of positive areas indicated a possible coexistence of PTGS2 and CD68 staining in some samples, while the correspondence of PTGS2 and CD163 was less evident (Figure 3a).

The quantification of cells, expressing these antigens in overlapping areas of equal extension, corroborated this observation (Figure 3b). The Pearson correlation coefficient was 0.422, (95% CI 0.229–0.582, p = 0.0000586) for CD68/PTGS2 and 0.316 (95% CI 0.110–0.496, p = 0.00324) for CD163/PTGS2.

To obtain a more specific quantification of the involvement of M1 and M2 macrophages in PTGS2 production in CRC, we also tested a multiplex IHC approach. Using consecutive destaining, stripping, and reprobing of the same tissue slices, we tested the CD68–iNOS–PTGS2 and the Arg1–MRC1–CD163–PTGS2 series (Figure 4).

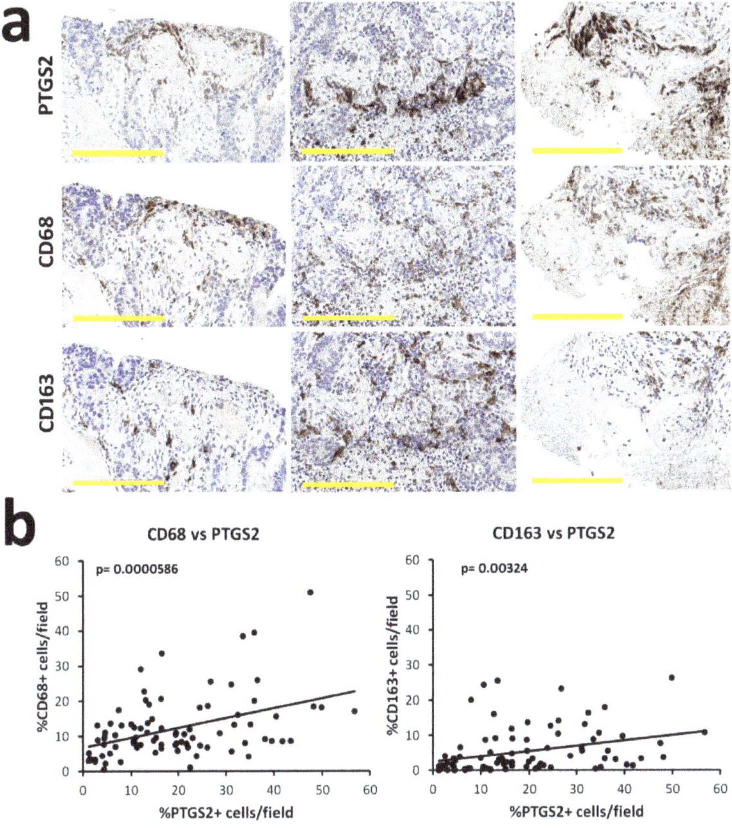

Figure 3. Analysis of PTGS2 influence on CD68 and CD163 macrophage populations on CRC serial sections: (**a**) Representative images of PTGS2, CD68, and CD163 staining on serial sections: PTGS2 can be found in areas enriched with CD68-positive macrophages, but it shows infrequent overlapping. Scale bar = 200 μm; (**b**) Digital pathology quantification of PTGS2, CD68, and CD163-positive cells on overlapping areas of serial sections and plot of linear regression. PTGS2 showed a partial relation with CD68-positive macrophages and a weak relation with the CD163 counterpart.

Multiplex analysis showed a complex picture with high variability among samples and different fields of the same sample. iNOS was strongly expressed by several non-macrophage cells, being the majority of iNOS positive signals in CRC tissues (Figure 4a). Few Arg1^{dim}-positive cells were detected, with minimal overlap with other M2 markers (Figure 4b and Supplementary Figure S1). A strong overlay was observed only between MRC1 and CD163 (Figure 4b), which was confirmed by co-localization analysis (Supplementary Figure S2). PTGS2 showed a limited co-localization with macrophages: the mean Pearson's coefficient for CD68-PTGS2 was 0.063, while the mean Manders' overlap coefficient, evaluating the extent of PTGS2 positivity in the CD68-positive area, was 0.237 (Figure 4c and Supplementary Figure S2). We also attempted a simplified representation of the co-localization of M1 and M2 single markers, assuming their expression only in macrophages (Figure 4c, right). This analysis suggested a mixed contribution of both M1 and M2 polarized cells in PTGS2 expression, with a possible prevalence of iNOS+ cells.

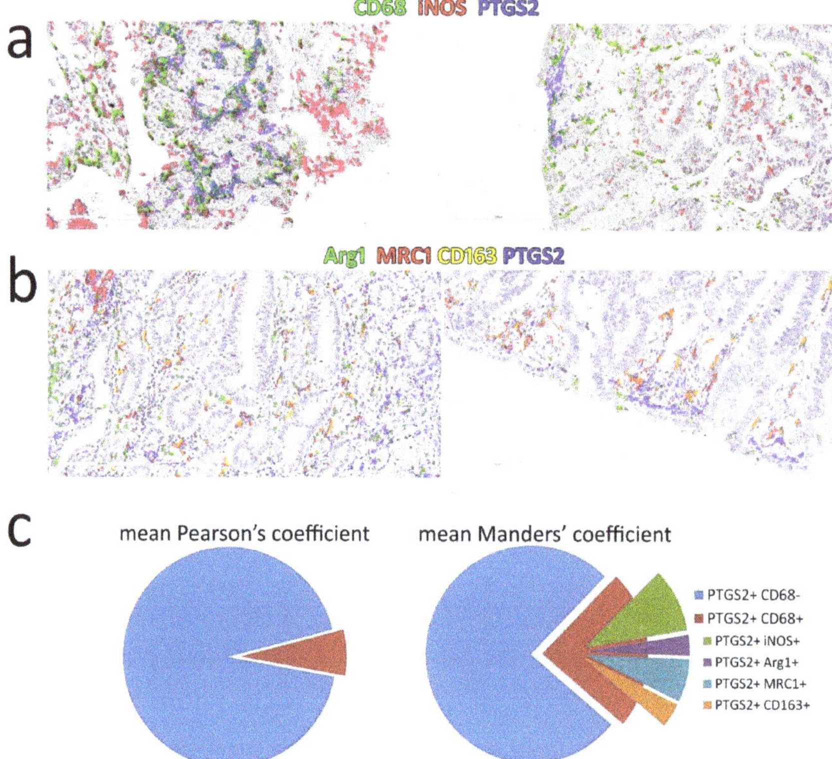

Figure 4. Multiplex IHC analysis of PTGS2 co-localization with macrophage markers CD68, iNOS, Arg1, MRC1/CD206, and CD163: (**a**) two examples of co-localization signals in the M1 series (CD68–iNOS–PTGS2), pseudocolors have been obtained overlaying the black&white mask of each marker on the first slide of the series; (**b**) two examples of co-localization signals in the M2 series (Arg1–MRC1–CD163–PTGS2); (**c**) left, mean Pearson's coefficients of PTGS2 and CD68 co-localization; right: a simplified representation of the mean overlay between PTGS2 and each marker using Manders' overlap coefficients. Slices indicate the extent of PTGS2 signal in the positive areas of each marker (complete analysis is shown in Supplementary Figure S2).

Thus, the expression of PTGS2 at the luminal surface of CRC, in non-tumor cells, did not apparently associate to a M1->M2 switch of macrophages. Moreover, macrophages did not appear as the main PTGS2-positive cell population in stromal areas.

Adegboyega et al. described subepithelial myofibroblasts as another source of PTGS2 in colorectal adenomas, mainly in luminal mucosal areas with damaged surface [21]. Sonoshita et al. observed the same localization of PTGS2 in the Apc$^{\Delta716}$ mouse model and in human adenomas, finding a strong co-localization of PTGS2 and vimentin, the marker of mesenchymal cells [22]. To verify the involvement of mesenchymal cells, we performed fluorescent double staining to co-localize PTGS2 and vimentin in multiple fields of representative CRC samples (Figure 5a).

Indeed, almost all PTGS2-positive cells corresponded to vimentin-positive cells, although stromal cells with the bright PTGS2 signal usually showed a lower signal for vimentin. Eliminating dim vimentin-positive signals by a color threshold, co-localization analysis showed a mean Pearson co-localization coefficient of 0.472 (Figure 5b; range 0.367–0.600, see Supplementary Figure S3a). Manders' overlap coefficients, evaluating the overlap of PTGS2 with vimentin and vice versa, showed that a high proportion of vimentin-positive PTGS2-negative mesenchymal cells was also present

(Supplementary Figure S3b). The mean of Manders' overlap coefficients evaluating the extent of PTGS2 signal in bright vimentin-positive areas was 0.301 (Figure 5c). The possible involvement of cancer-associated fibroblasts (CAF) as a prevalent non-tumor source of PTGS2 in CRC led us to analyze in vitro the possible inducers of PTGS2 in primary colorectal CAF.

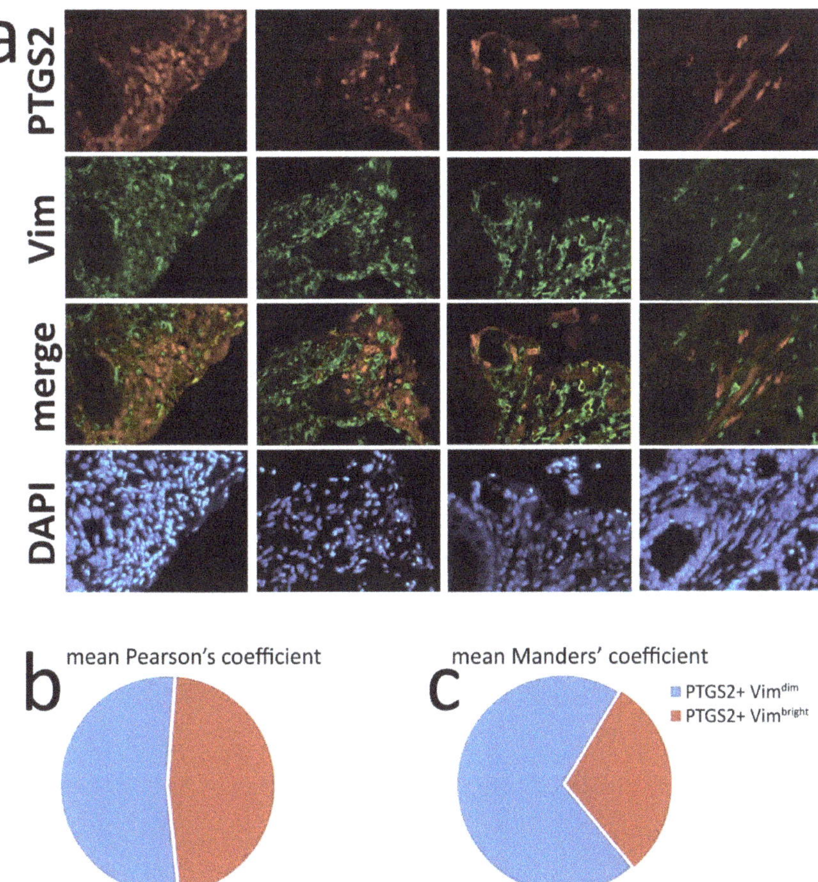

Figure 5. Immunofluorescent double staining of PTGS2 and vimentin indicates cancer-associated fibroblasts as a major source of PTGS2: (**a**) Representative images of PTGS2 and vimentin double immunofluorescence in four CRC. PTGS2 positivity almost invariably corresponded to vimentin positive cells, although bright PTGS2 staining frequently corresponded to dim vimentin staining and vice versa; (**b**) mean PTGS2–vimentin co-localization evaluated by Pearson's coefficient after the subtraction of dim vimentin positivity; (**c**) mean Manders' overlap coefficient evaluating the extent of PTGS2 signal in bright vimentin-positive areas (complete analysis is shown in Supplementary Figure S3).

3.3. In Vitro, IL1β is a Powerful Inducers of PTGS2 in CAF and Correlates to gPTGS2 Levels in Tissue Lysates

CAF primary cells [15] were stimulated for 24 h with IL1β, a known inducer of PTGS2 expression by NFkB and AP1 activation [23]. Other factors linked to inflammatory/trophic CRC progression (IL8/CXCL8, GROβ/CXCL2, as agonists of G-protein coupled receptors; PGE2 produced by PTGS2 activation; EGF as the prototype EGFR tyrosine-kinase agonist) were also tested as potential PTGS2 inducers [1,24–26]. The analysis for gPTGS2 expression by Western blot showed a powerful induction of gPTGS2 by IL1β, and weak responses elicited by PGE2 and EGF (Figure 6a).

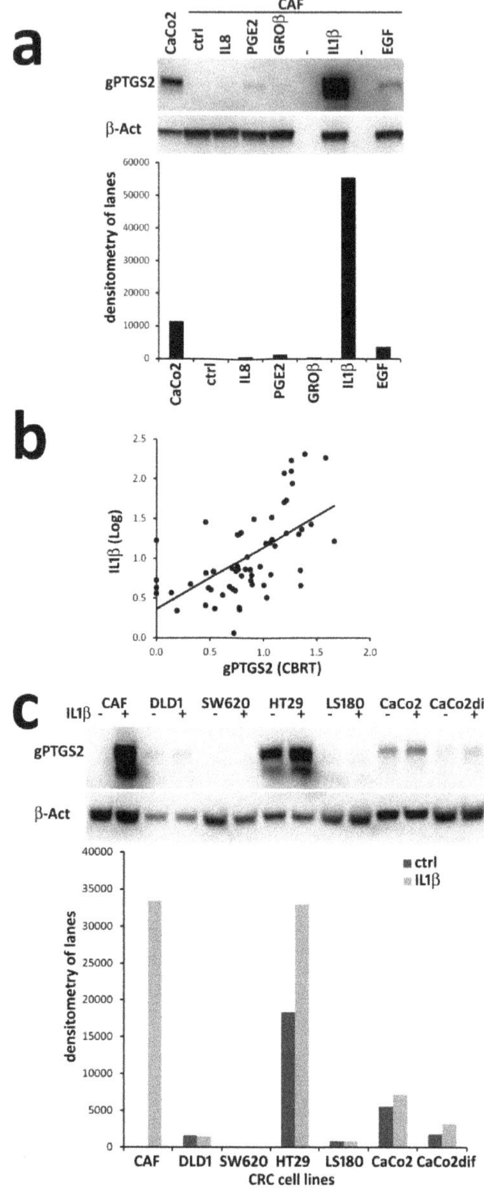

Figure 6. IL1β strongly upregulates PTGS2 expression in cancer-associated fibroblasts (CAF) and correlates to gPTGS2 levels in tissue lysates: (**a**) The well-known PTGS2 inducer IL1β and other four cytokines involved in CRC (CXCL8/IL8, PGE2, CXCL2/GROβ, and EGF; 24 h treatment) were tested as possible inducers of PTGS2 expression in primary CRC CAF. Only IL1β induced a powerful response. Densitometric quantifications of WB lanes are plotted; (**b**) Linear regression of PTGS2 and IL1β levels in CRC tissue lysates (60 samples). Pearson's correlation coefficient of gPTGS2 (CBRT = cubic root) vs. IL1β (Log) levels was 0.593 (p = 0.00000609; (**c**) IL1β (24 h treatment) has a reduced ability to modify PTGS2 basal expression in CRC cell lines. Densitometric quantifications of Western blot lanes are plotted. WB were run as experimental duplicates, with similar results (see Supplementary Figure S4).

IL1β involvement in the induction of gPTGS2 in our cohort was verified by the quantification of its levels in 60 tissue lysates (Figure 6b). Pearson correlation coefficient of gPTGS2 versus IL1β levels was 0.593 (p = 0.00000609), indicating a strong link between IL1β and gPGTS2 levels in more than half of the samples. Western blot analysis of CRC cell lines stimulated by IL1β revealed a possible explanation for the existence of samples with a low gPTGS2/IL1β correlation. CRC cell lines showed high or low basal levels of gPTGS2 with a limited response to IL1β stimulation compared to CAF (Figure 6c), even at 10× increased doses of IL1β (Supplementary Figure S4). Thus, the prevalence of stroma, or tumor-derived gPTGS2, could mirror the strong or weak response to the presence of IL1β in CRC tissues.

3.4. Effects of Stromal PTGS2 on Patients Prognosis

In our cohort, stromal PTGS2 affected patients OS according to low, medium, or high IHC positivity (Figure 7a). In particular, intermediate levels of stromal PTGS2 (n = 22) were apparently associated to a better prognosis, while high PTGS2 (n = 6) showed a negative outcome compared to low/null PTGS2-expressing CRC.

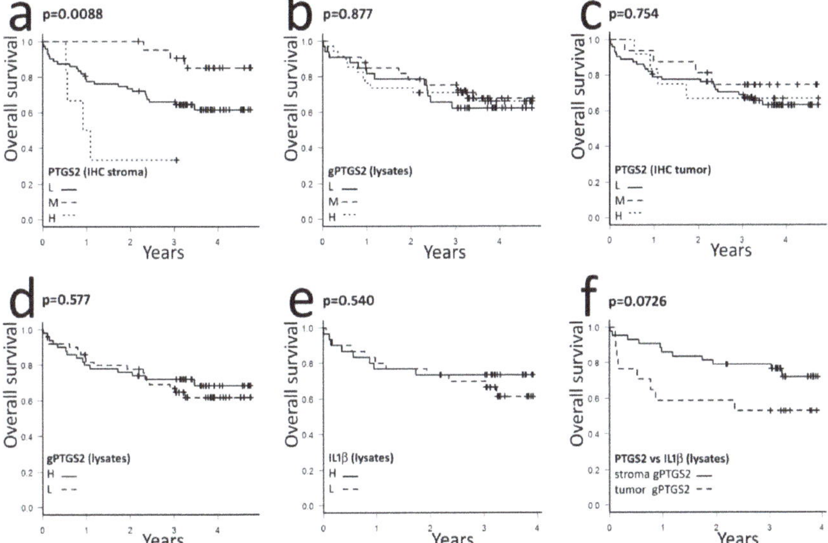

Figure 7. Effects of PTGS2 localization on patients overall survival: (**a**) Kaplan–Meier analysis of patients' overall survival according to PTGS2 levels quantified by IHC in stromal cells. L = low, M = medium, or H = high PTGS2 expression; (**b**) Kaplan–Meier analysis of patients' overall survival according to gPTGS2 levels quantified in tissue lysates by WB. L=low, M=medium, or H=high gPTGS2 expression; (**c**) Kaplan–Meier analysis of patients overall survival according to PTGS2 levels quantified by IHC in tumor epithelial cells. L=low, M=medium, or H=high PTGS2 expression: (**d–e**) gPTGS2 and IL1β levels detected in tissue lysates by WB were dichotomized (low vs. high) using the median value of data as a threshold. Kaplan–Meier analysis of patients' overall survival against low or high gPTGS2 (d) or IL1β (e) levels was not affected by this dichotomization; (**f**) Dichotomization shown in panels d–e was used to separate tumors into two new classes: the first with a directly proportional regulation of gPTGS2 and IL1β (gPTGS2^{low}IL1β^{low} + gPTGS2^{high}IL1β^{high} = stroma gPTGS2 enriched, continuous line), the other with an independent regulation of these markers (gPTGS2^{low}IL1β^{high} + gPTGS2^{high}IL1β^{low} = tumor gPTGS2 enriched, dashed line). The Kaplan–Meier analysis of patients' overall survival based on this subdivision is shown.

The exclusion of stage IV tumors from Kaplan–Meier analysis did not modify the prognostic potential of stromal PTGS2 expression (Supplementary Figure S5), reducing a potential bias on OS evaluation linked to the surgical or therapeutic treatment of this group of patients. Neither gPTGS2 total levels quantified in tissue lysates nor tumor-associated PTGS2 scored by IHC influenced patients OS in our cohort (Figure 7b,c).

Our observations suggested a possible subgrouping of CRC on the basis of stromal, or tumor-derived PTGS2, the former modulated by IL1β, with a prognostic significance, the latter relatively independent. We asked if we could identify these two categories of tumors on the basis of gPTGS2 and IL1β levels in tissue lysates and if they could influence patients prognosis. When dichotomized by the median value, PTGS2 showed no relation with the main pathological data (Table 1), but it maintained a significant relation with IL1β (the database with pathological data, PTGS2, and IL1β quantification is available in Supplementary data S1).

Table 1. Clinical and molecular features of CRC according to median PTGS2 levels.

Clinicopathologic Parameters	All	PTGS2 < Median	PTGS2 > Median	*p*
TOTAL	**100**	**50**	**50**	-
SEX				
Male	**49**	26	23	*0.689*
Female	**51**	24	27	
AGE (surgery)				
≤70	**47**	22	28	*0.317*
>70	**53**	28	22	
TUMOR LOCATION *				
Proximal colon	**43**	21	22	
Distal colon	**36**	18	18	*0.965*
Rectum	**21**	11	10	
STAGE (UICC-2009)				
I	**14**	7	7	
II	**34**	16	18	*0.974*
III	**39**	20	19	
IV	**13**	7	6	
TUMOR GRADE				
well	**5**	2	4	
moderate	**75**	40	35	*0.479*
poor	**19**	8	11	
JASS SCORE				
I	**17**	7	10	
II	**14**	6	8	*0.719*
III	**32**	18	14	
IV	**37**	19	18	
PERINEURAL INVASION				
PNI 0	**69**	37	32	*0.387*
PNI 1	**31**	13	18	
MICROSAT. INSTAB. **				
MSS	**72**	40	32	*0.158*
MSI	**15**	5	10	
IL1β ***				
IL1β<median	**30**	21	8	*0.002*
IL1β>median	**30**	10	22	

* Proximal colon includes cecum to transverse colon; distal colon includes splenic flexure to sigmoid colon. ** Data available for 87 samples. *** Data available for 60 samples. NOTE: 2 × 2 tables were tested by Fisher's exact test, other parameters by Chi-square test.

We used gPTGS2 and IL1β dichotomization by the median of data and defined as stroma-derived gPTGS2 the cases with low + low or high + high gPTGS2 and IL1β levels (= directly proportional IL1β and gPTGS2 levels). On the contrary, tumor-derived gPTGS2 was defined in cases with low + high or high + low gPTGS2 and IL1β levels (= independence of IL1β and gPTGS2 expression). Dichotomized gPTGS2 or IL1β levels, individually assessed by Kaplan–Meier analysis, did not influence patients' OS, showing almost superimposable curves (Figure 7d,e). On the contrary, an immediate divergence of curves was observed when CRC with putative stroma-derived PTGS2 were compared to CRC with putative tumor-derived PTGS2 (Figure 7f). This categorization suggested a different involvement of stroma-expressed versus tumor-expressed PTGS2 in patient outcomes, although it did not reach statistical significance.

4. Discussion

Several epidemiological studies suggested a favorable outcome for patients taking NSAIDS after CRC diagnosis/surgery [5–7]. An ideal approach for PTGS2 targeting would be the identification of CRC patients that could benefit from the inclusion of PTGS2 inhibitors in the adjuvant setting. The influence of PTGS2 expression on patient prognosis could be a possible criterion, but the quantification of PTGS2 by IHC only in CRC tumor cells showed variable results [8–11]. We attempted to overcome this limitation quantifying gPTGS2 levels in CRC lysates, comparing them with IHC-scored, tumor, or stroma-derived PTGS2.

Our approach shows that gPTGS2 levels can be quantified in CRC lysates with high sensitivity and specificity. gPTGS2 levels partially correlated with both tumor-associated and stroma-associated PTGS2 detected by IHC, indicating that gPTGS2 expression is not restricted to tumor cells as hypothesized before. Stroma-associated PTGS2 showed an almost exclusive luminal distribution, as already observed in adenomas [21,22], suggesting a homeostatic role in preserving the mucosal barrier. Macrophages did not appear as a major PTGS2-positive cell population, although minor populations of macrophages expressing M1 (iNOS) or M2 (MRC1, CD163) markers co-localized with the PTGS2 signal. The slight prevalence of the PTGS2 signal in iNOS-positive macrophages would be in accordance with the preferential expression of PTGS2 by M1-polarized macrophages observed in vitro [27]. Multiplex IHC analysis showed that only MRC1 and CD163 were frequently associated, while Arg1 expression was rare and iNOS was expressed by several different cells without relation to CD68. While macrophages could directly contribute to PTGS2 expression in CRC, their specific ability to produce IL1β [28] would be a major mechanism of PTGS2 amplification by the paracrine stimulation of bystander fibroblasts. Indeed, Cui et al. showed that most IL1β-expressing cells localize in the stroma of CRC (median 19.2 cells/high power field, hpf), while positive epithelial tumor cells are rare (median 0.4 cells/hpf) [29]. Our in vitro data and CRC lysates analysis sustain the hypothesis that gPTGS2 expression could be mediated by IL1β, preferentially targeting CAF. Accordingly, Cui et al. showed a higher positivity for IL1R1 in the stroma (median 11.0 cells/hpf) than in tumor epithelial cells (median 0.7 cells/hpf) [29]. IL1β is neo-synthesized and activated by the inflammasome only in the presence of microbial components or tissue damage [28], sustaining the plausible role of luminal, stromal PTGS2 for the homeostatic rescue of the epithelial barrier. Indeed, several bacterial species have been associated to CRC [30], and *Streptococcus gallolyticus* can infect colorectal tumors and has been linked to local IL1β and PTGS2 expression [31]. According to these data, stromal PTGS2 expressed in the luminal area could exert a protective role for the patient, not necessarily influencing CRC progression. Unfortunately, we could not evaluate this hypothesis as only overall survival data were available for our cohort, without information about the cause of death.

Tumor-associated PTGS2 did not apparently affect patients' overall survival, and the quantification of specific gPTGS2 levels in tissue lysates could not discriminate patient outcome. Tumor epithelial PTGS2 could be less controlled by physiologic stimuli as some oncogenic mechanisms can affect its expression; for example, PTGS2 is frequently downregulated in MSI CRC, while PIK3CA mutation could mediate PTGS2 activity [32–34]. When gPTGS2 and IL1β levels were used to discriminate those

CRC depending on IL1β for gPTGS2 expression (enriched in stromal gPTGS2) from tumors with an independent gPTGS2 expression (enriched in tumor gPTGS2), survival curves showed an immediate dichotomy. While this datum has no significant prognostic implications, it suggests distinct roles for stromal and tumor PTGS2.

5. Conclusion

Our study suggests an unpredicted association of stromal PTGS2 with patients' prognosis that limits the use of total gPTGS2 quantification in CRC samples lysates for predictive purposes. Due to the possible positive influence on patient OS of an intermediate PTGS2 expression in the luminal tumor stroma, we propose further validation of this marker on larger cohorts.

Supplementary Materials: The following are available online at http://www.mdpi.com/2073-4409/9/3/683/s1, Table S1: PTGS2/COX2 expression in CRC (literature); Supplementary method S1: anti-PTGS2 antibody selection; Figure S1: Example of multiplex IHC and method for overlay computation; Figure S2: Analysis of 1:1 colocalization of M1 and M2 macrophage markers with PTGS2 by the JACoP plug-in; Figure S3: JACop colocalization analysis of PTGS2 and Vimentin fluorescent staining; Figure S4: Confirm of Western blot data shown in Figure 6; Figure S5: Effects of PTGS2 localization on patients overall survival (stage IV CRC excluded); Supplemetary data S1: database with pathological data and PTGS2 and IL1β quantification (excel file).

Author Contributions: Conceptualization, R.V. and R.B.; Data curation, R.V., D.C. and R.B.; Formal analysis, R.B.; Funding acquisition, R.V. and A.P.; Investigation, R.V., D.C., R.A. and F.P.; Methodology, R.V., D.C. and R.B.; Project administration, R.V., S.S., L.M., E.R., A.P. and R.B.; Resources, S.C., S.S., G.C.P., M.B., S.Z., F.G., L.M., S.M. and E.R.; Supervision, A.P. and R.B.; Validation, R.V., D.C., F.G., L.M. and R.B.; Visualization, R.B.; Writing—original draft, R.B.; Writing—review & editing, R.V., F.G., L.M., N.F., F.T., M.C.M., A.P. and R.B. All authors have read and agreed to the published version of the manuscript.

Funding: This research was funded by Ministero della Salute ricerca finalizzata, progetto giovani ricercatori [R.V. GR-2013-02356568]; Ministero della salute fondo 5x1000 Enti della Ricerca Sanitaria (2014, 2015, 2016); Ministero della Salute ricerca corrente (2018–2021); Associazione Italiana per la Ricerca sul Cancro (AIRC) [A.P. IG21648].

Acknowledgments: We wish to acknowledge Francesca Sarocchi, Rossella Ponte, Paola Calamaro and Damiana LaTorre for their support in sample collection. We wish to acknowledge Simona Pigozzi for the excellent technical support in IHC. We also acknowledge Maria Raffaella Zocchi for critical revision of the manuscript.

Conflicts of Interest: The authors declare no conflict of interest. The funders had no role in the design of the study; in the collection, analyses, or interpretation of data; in the writing of the manuscript, or in the decision to publish the results.

References

1. Benelli, R.; Venè, R.; Ferrari, N. Prostaglandin-endoperoxide synthase 2 (cyclooxygenase-2), a complex target for colorectal cancer prevention and therapy. *Transl. Res.* **2018**, *196*, 42–61. [CrossRef] [PubMed]

2. Jaén, R.I.; Prieto, P.; Casado, M.; Martín-Sanz, P.; Boscá, L. Post-translational modifications of prostaglandin-endoperoxide synthase 2 in colorectal cancer: An update. *World J. Gastroenterol.* **2018**, *24*, 5454–5461. [CrossRef] [PubMed]

3. Arber, N.; Eagle, C.J.; Spicak, J.; Rácz, I.; Dite, P.; Hajer, J.; Zavoral, M.; Lechuga, M.J.; Gerletti, P.; Tang, J.; et al. Celecoxib for the Prevention of Colorectal Adenomatous Polyps. *N. Engl. J. Med.* **2006**, *355*, 885–895. [CrossRef]

4. Bertagnolli, M.M.; Eagle, C.J.; Zauber, A.G.; Redston, M.; Solomon, S.D.; Kim, K.; Tang, J.; Rosenstein, R.B.; Wittes, J.; Corle, D.; et al. Celecoxib for the Prevention of Sporadic Colorectal Adenomas. *N. Engl. J. Med.* **2006**, *355*, 873–884. [CrossRef] [PubMed]

5. Ng, K.; Meyerhardt, J.A.; Chan, A.T.; Sato, K.; Chan, J.A.; Niedzwiecki, D.; Saltz, L.B.; Mayer, R.J.; Benson, A.B.; Schaefer, P.L.; et al. Aspirin and COX-2 inhibitor use in patients with stage III colon cancer. *J. Natl. Cancer Inst.* **2015**, *107*, 345. [CrossRef] [PubMed]

6. Friis, S.; Riis, A.H.; Erichsen, R.; Baron, J.A.; Sørensen, H.T. Low-Dose Aspirin or Nonsteroidal Anti-inflammatory Drug Use and Colorectal Cancer Risk: A Population-Based, Case-Control Study. *Ann. Intern. Med.* **2015**, *163*, 347–355. [CrossRef] [PubMed]

7.   Hua, X.; Adams, S.V.; Phipps, A.I.; Cohen, S.A.; Burnett-Hartman, A.; Hardikar, S.; Newcomb, P.A. Pre- and Post-Diagnostic Non-Steroidal Anti-Inflammatory Drug Use and Colorectal Cancer Survival in Seattle Colon Cancer Family Registry. *Cancer Epidemiol. Biomarkers Prev.* **2016**, *25*, 559. [CrossRef]

8.   Midgley, R.S.; McConkey, C.C.; Johnstone, E.C.; Dunn, J.A.; Smith, J.L.; Grumett, S.A.; Julier, P.; Iveson, C.; Yanagisawa, Y.; Warren, B.; et al. Phase III randomized trial assessing rofecoxib in the adjuvant setting of colorectal cancer: Final results of the VICTOR trial. *J. Clin. Oncol.* **2010**, *28*, 4575–4580. [CrossRef]

9.   Fux, R.; Schwab, M.; Thon, K.-P.; Gleiter, C.H.; Fritz, P. Cyclooxygenase-2 Expression in Human Colorectal Cancer Is Unrelated to Overall Patient Survival. *Clin. Cancer Res.* **2005**, *11*, 4754–4760. [CrossRef]

10. Ogino, S.; Kirkner, G.J.; Nosho, K.; Irahara, N.; Kure, S.; Shima, K.; Hazra, A.; Chan, A.T.; Dehari, R.; Giovannucci, E.L.; et al. Cyclooxygenase-2 expression is an independent predictor of poor prognosis in colon cancer. *Clin. Cancer Res.* **2008**, *14*, 8221–8227. [CrossRef]

11. Soumaoro, L.T.; Uetake, H.; Higuchi, T.; Takagi, Y.; Enomoto, M.; Sugihara, K. Cyclooxygenase-2 expression: A significant prognostic indicator for patients with colorectal cancer. *Clin. Cancer Res.* **2004**, *10*, 8465–8471. [CrossRef] [PubMed]

12. Conti, J.; Thomas, G. The Role of Tumour Stroma in Colorectal Cancer Invasion and Metastasis. *Cancers* **2011**, *3*, 2160–2168. [CrossRef] [PubMed]

13. Mei, Z.; Liu, Y.; Liu, C.; Cui, A.; Liang, Z.; Wang, G.; Peng, H.; Cui, L.; Li, C. Tumour-infiltrating inflammation and prognosis in colorectal cancer: Systematic review and meta-analysis. *Br. J. Cancer* **2014**, *110*, 1595–1605. [CrossRef] [PubMed]

14. Roseweir, A.K.; Park, J.H.; Hoorn, S.; Powell, A.G.; Roxburgh, C.S.; McMillan, D.C.; Horgan, P.G.; Vermeulen, L.; Edwards, J. Abstract 4615: Phenotypic subtypes successfully stratify prognosis of patient with colorectal cancer: A step towards precision medicine. *Cancer Res.* **2018**, *78*, 4615.

15. Benelli, R.; Venè, R.; Minghelli, S.; Carlone, S.; Gatteschi, B.; Ferrari, N. Celecoxib induces proliferation and Amphiregulin production in colon subepithelial myofibroblasts, activating erk1-2 signaling in synergy with EGFR. *Cancer Lett.* **2013**, *328*, 73–82. [CrossRef]

16. Chapple, K.S.; Cartwright, E.J.; Hawcroft, G.; Tisbury, A.; Bonifer, C.; Scott, N.; Windsor, A.C.; Guillou, P.J.; Markham, A.F.; Coletta, P.L.; et al. Localization of cyclooxygenase-2 in human sporadic colorectal adenomas. *Am. J. Pathol.* **2000**, *156*, 545–553. [CrossRef]

17. Bamba, H.; Ota, S.; Kato, A.; Adachi, A.; Itoyama, S.; Matsuzaki, F. High expression of cyclooxygenase-2 in macrophages of human colonic adenoma. *Int. J. Cancer* **1999**, *83*, 470–475. [CrossRef]

18. Heusinkveld, M.; van der Burg, S.H. Identification and manipulation of tumor associated macrophages in human cancers. *J. Transl. Med.* **2011**, *9*, 216. [CrossRef]

19. Nakanishi, Y.; Nakatsuji, M.; Seno, H.; Ishizu, S.; Akitake-Kawano, R.; Kanda, K.; Ueo, T.; Komekado, H.; Kawada, M.; Minami, M.; et al. COX-2 inhibition alters the phenotype of tumor-associated macrophages from M2 to M1 in ApcMin/+ mouse polyps. *Carcinogenesis* **2011**, *32*, 1333–1339. [CrossRef]

20. Sheppe, A.E.F.; Kummari, E.; Walker, A.; Richards, A.; Hui, W.W.; Lee, J.H.; Mangum, L.; Borazjani, A.; Ross, M.K.; Edelmann, M.J. PGE2 Augments Inflammasome Activation and M1 Polarization in Macrophages Infected With Salmonella Typhimurium and Yersinia enterocolitica. *Front. Microbiol* **2018**, *9*, 2447. [CrossRef]

21. Adegboyega, P.A.; Ololade, O.; Saada, J.; Mifflin, R.; Di Mari, J.F.; Powell, D.W. Subepithelial myofibroblasts express cyclooxygenase-2 in colorectal tubular adenomas. *Clin. Cancer Res.* **2004**, *10*, 5870–5879. [CrossRef] [PubMed]

22. Sonoshita, M.; Takaku, K.; Oshima, M.; Sugihara, K.; Taketo, M.M. Cyclooxygenase-2 Expression in Fibroblasts and Endothelial Cells of Intestinal Polyps. *Cancer Res.* **2002**, *62*, 6846–6849. [PubMed]

23. Ramsay, R.G.; Ciznadija, D.; Vanevski, M.; Mantamadiotis, T. Transcriptional regulation of cyclo-oxygenase expression: Three pillars of control. *Int. J. Immunopathol. Pharmacol* **2003**, *16*, 59–67. [PubMed]

24. Ning, Y.; Lenz, H.-J. Targeting IL-8 in colorectal cancer. *Expert Opin. Ther. Tar.* **2012**, *16*, 491–497. [CrossRef] [PubMed]

25. Wang, G.; Huang, J.; Zhu, H.; Ju, S.; Wang, H.; Wang, X. Overexpression of GRO-β is associated with an unfavorable outcome in colorectal cancer. *Oncol. Lett.* **2016**, *11*, 2391–2397. [CrossRef] [PubMed]

26. Wang, D.; Xia, D.; DuBois, R.N. The Crosstalk of PTGS2 and EGF Signaling Pathways in Colorectal Cancer. *Cancers* **2011**, *3*, 3894–3908. [CrossRef] [PubMed]

27. Martinez, F.O.; Gordon, S.; Locati, M.; Mantovani, A. Transcriptional Profiling of the Human Monocyte-to-Macrophage Differentiation and Polarization: New Molecules and Patterns of Gene Expression. *J. Immunol.* **2006**, *177*, 7303–7311. [CrossRef]
28. Voronov, E.; Apte, R.N. IL-1 in Colon Inflammation, Colon Carcinogenesis and Invasiveness of Colon Cancer. *Cancer Microenviron.* **2015**, *8*, 187–200. [CrossRef]
29. Cui, G.; Yuan, A.; Sun, Z.; Zheng, W.; Pang, Z. IL-1β/IL-6 network in the tumor microenvironment of human colorectal cancer. *Pathol. Res. Pract.* **2018**, *214*, 986–992. [CrossRef]
30. Dahmus, J.D.; Kotler, D.L.; Kastenberg, D.M.; Kistler, C.A. The gut microbiome and colorectal cancer: A review of bacterial pathogenesis. *J. Gastrointest. Oncol.* **2018**, *9*, 769–777. [CrossRef]
31. Abdulamir, A.S.; Hafidh, R.R.; Bakar, F.A. Molecular detection, quantification, and isolation of Streptococcus gallolyticus bacteria colonizing colorectal tumors: Inflammation-driven potential of carcinogenesis via IL-1, COX-2, and IL-8. *Mol. Cancer* **2010**, *9*, 249. [CrossRef] [PubMed]
32. Castells, A.; Payá, A.; Alenda, C.; Rodríguez-Moranta, F.; Agrelo, R.; Andreu, M.; Piñol, V.; Castellví-Bel, S.; Jover, R.; Llor, X.; et al. Cyclooxygenase 2 expression in colorectal cancer with DNA mismatch repair deficiency. *Clin. Cancer Res.* **2006**, *12*, 1686–1692. [CrossRef] [PubMed]
33. Karnes, W.E.; Shattuck-Brandt, R.; Burgart, L.J.; DuBois, R.N.; Tester, D.J.; Cunningham, J.M.; Kim, C.Y.; McDonnell, S.K.; Schaid, D.J.; Thibodeau, S.N. Reduced COX-2 protein in colorectal cancer with defective mismatch repair. *Cancer Res.* **1998**, *58*, 5473–5477. [PubMed]
34. Domingo, E.; Church, D.N.; Sieber, O.; Ramamoorthy, R.; Yanagisawa, Y.; Johnstone, E.; Davidson, B.; Kerr, D.J.; Tomlinson, I.P.M.; Midgley, R. Evaluation of PIK3CA Mutation As a Predictor of Benefit From Nonsteroidal Anti-Inflammatory Drug Therapy in Colorectal Cancer. *JCO* **2013**, *31*, 4297–4305. [CrossRef]

Article

miR-9-Mediated Inhibition of *EFEMP1* Contributes to the Acquisition of Pro-Tumoral Properties in Normal Fibroblasts

Giulia Cosentino [1], Sandra Romero-Cordoba [1,2], Ilaria Plantamura [1], Alessandra Cataldo [1,*,†] and Marilena V. Iorio [1,3,*,†]

1   Molecular Targeting Unit, Research Department, Fondazione IRCCS Istituto Nazionale dei Tumori, 20133 Milan, Italy; giulia.cosentino@istitutotumori.mi.it (G.C.); sandra.romeroc@incmnsz.mx (S.R.-C.); ilaria.plantamura@istitutotumori.mi.it (I.P.)
2   Biochemistry Department, Instituto Nacional de Ciencias Médicas y Nutriciòn Salvador Zubirán, Mexico City 14080, Mexico
3   Istituto FIRC Oncologia Molecolare (IFOM), 20139 Milan, Italy
*   Correspondence: alessandra.cataldo@istitutotumori.mi.it (A.C.); marilena.iorio@istitutotumori.mi.it (M.V.I.); Tel.: +39-022-390-5134 (M.V.I.)
†   These authors equally contributed to this work.

Received: 20 July 2020; Accepted: 18 September 2020; Published: 22 September 2020

Abstract: Tumor growth and invasion occurs through a dynamic interaction between cancer and stromal cells, which support an aggressive niche. MicroRNAs are thought to act as tumor messengers to "corrupt" stromal cells. We previously demonstrated that miR-9, a known metastamiR, is released by triple negative breast cancer (TNBC) cells to enhance the transition of normal fibroblasts (NFs) into cancer-associated fibroblast (CAF)-like cells. EGF containing fibulin extracellular matrix protein 1 (*EFEMP1*), which encodes for the ECM glycoprotein fibulin-3, emerged as a miR-9 putative target upon miRNA's exogenous upmodulation in NFs. Here we explored the impact of *EFEMP1* downmodulation on fibroblast's acquisition of CAF-like features, and how this phenotype influences neoplastic cells to gain chemoresistance. Indeed, upon miR-9 overexpression in NFs, *EFEMP1* resulted downmodulated, both at RNA and protein levels. The luciferase reporter assay showed that miR-9 directly targets *EFEMP1* and its silencing recapitulates miR-9-induced pro-tumoral phenotype in fibroblasts. In particular, *EFEMP1* siRNA-transfected (si-*EFEMP1*) fibroblasts have an increased ability to migrate and invade. Moreover, TNBC cells conditioned with the supernatant of NFs transfected with miR-9 or si-*EFEMP1* became more resistant to cisplatin. Overall, our results demonstrate that miR-9/*EFEMP1* axis is crucial for the conversion of NFs to CAF-like cells under TNBC signaling.

Keywords: tumor microenvironment; triple-negative breast cancer; cancer-associated fibroblasts; *EFEMP1*; miRNA; miR-9; chemoresistance

---

1. Introduction

The physiological role of stromal cells like fibroblasts, endothelial cells, adipocytes and immune cells is to sustain and shield epithelial cells from harm [1]. Breast cancer, as other solid tumors, must engage stromal cells in an aberrant cross-talk in order to grow, invade the neighboring tissues, and migrate to distant sites [2]. For example, "corrupted" fibroblasts, the so-called cancer-associated fibroblasts (CAFs), actively secrete pro-tumor factors like growth factors, cytokines and chemokines, remodel the extracellular matrix (ECM) to favor tumor cell motility and, eventually, mediate resistance to anticancer drugs [3–5]. CAFs are also able to affect the behavior of the other stromal cells, for instance by releasing pro-inflammatory chemokines and pro-angiogenic factors that facilitate the immune and endothelial cell recruitment at the tumor site and the polarization toward a malignant phenotype [6–8].

Triple-negative breast cancer, a highly aggressive malignancy, is thought to have a unique microenvironment, distinct from other breast cancer subtypes, which might significantly impact on the progression of these malignances [9].

An increasing body of evidence supports the involvement of microRNAs (miRNAs) in the interaction between tumor and stroma. Indeed miRNAs, small non-coding RNAs involved in post-transcriptional gene regulation, have been proven to act as "messages" to induce the acquisition of malignant traits in stromal cells [10]. Accordingly, in our previous work by Baroni S et al., we demonstrated that TNBC cells are able to induce the acquisition of CAF-like properties in NFs by releasing the known breast metastamiR miR-9, packaged into exosomes [11,12]. We also showed that these CAF-like cells can increase, in turn, tumor cell aggressiveness. Gene expression profile of miR-9 overexpressing NFs revealed *EFEMP1*, collagen type1 alpha1 (COL1A1) and matrix metalloproteinase-1 (MMP1), as the most significantly modulated genes, being the first two transcripts predicted miR-9 targets. These molecules were selected for further analyses since they are known to be involved in the crucial pathways of ECM synthesis and remodelling. However, since only *EFEMP1* downmodulation was validated in public datasets comparing tumor vs normal stroma of breast cancer patients [11], we decided to focus our efforts on studying *EFEMP1* contribution to the observed phenotype.

EFEMP1 encodes for the ECM glycoprotein fibulin-3, which participates in maintaining the integrity of the stroma linking elastic fibres to basement membranes [13,14]. Interestingly, in 2015 Tian H et al. identified fibulin-3 as a novel TGF-β pathway inhibitor in breast cancer microenvironment, interfering with tumor progression [15]. Here we focus on validating *EFEMP1* targeting by miR-9 in fibroblasts and explore the contribution of this modulation to the acquisition of CAF-like features, such as cell motility and induction of chemoresistance in TNBC cells.

2. Materials and Methods

2.1. In-Silico Analysis to Define Caf and NF EFEMP1 Expression Portraits

Normalized gene expression profiles of GSE20086, GSE80035 and GSE37614 were downloaded from Geo omnibus. Genes were annotated with biomaRt package from Bioconductor in R environment [16]. Duplicated probes for a same gene were collapsed by selecting the one with the highest interquartile range for Affymetrix profiling, while the probe with the highest value was selected for further analyses on Illumina profiles. Plots were performed with ggplot. Wilcoxon test was applied to define differential expression on R.

2.2. Cell Culture and Primary Fibroblasts Isolation

Immortalized normal fibroblasts, HEK-293T and MDA-MB-468 cell lines were purchased from ATCC (Rockville, MD, USA). NFs were cultured in FGM-2 medium with 10% FBS, HEK-293T and MDA-MB-468 in DMEM with 10% FBS and maintained at 37 °C under 5% CO_2. MycoAlert Mycoplasma Detection Kit (Lonza, Basel, Switzerland) was used to assure a negative mycoplasma status in cultured cells before experiments were started. Primary NFs and CAFs were isolated from specimen belonging to TNBC patient who underwent surgery at Fondazione IRCCS Istituto Nazionale dei Tumori of Milan (INT) and who signed an informed consent to donate the leftover tissue after diagnosis to INT for research. The INT Ethic Committee authorized the use of these samples for the project "Tumor-microenvironment related changes as new tools for early detection and assessment of high-risk disease" on January 24th 2012. RNA from these samples was isolated as previously described [11].

2.3. MiRNA Mimics and siRNA Transient Transfection

MiR-9 overexpression was performed using a chemically synthesized miRNA mimic (Catalog number AM17100, Assay ID PM10022, Thermo Fisher Scientific, Waltham, MA, USA) at a final concentration of 100 nM. A Silencer® Select Pre-Designed siRNA (Catalog number AM16708, Assay ID 14094 Thermo Fisher Scientific, Waltham, MA, USA) was purchased to perform *EFEMP1* silencing, using a final concentration of 50 nM. Lipofectamine 2000 was used as transfection reagent in Optimem medium (Gibco, Thermo Fisher Scientific, Waltham, MA, USA), which was replaced with standard medium after 6 h.

2.4. Cloning and Mutagenesis

EFEMP1 3′UTR was cloned into pmirGLO vector plasmid (Promega, Medison, WI, USA), designed to perform luciferase reporter assay and carrying β-lactamase coding region (Ampicillin resistance). *EFEMP1* 3′UTR sequence to be cloned was amplified by PCR using ThermoScientific Phusion Hot Start High-Fidelity DNA polymerase kit (Thermo Fisher Scientific, Waltham, MA, USA). Primer sequences are reported in Table 1. Plasmid vector and insert were first digested with NheI and XbaI restriction enzymes (New England Biolabs, Ipswich, MA, USA) through incubation for 1h at 37 °C. The digested products were purified with Gel/PCR DNA Fragments Extraction kit, dephosphorylated with rAPid Alkaline Phosphatase kit (Roche, Basel, Switzerland) through incubation at 37 °C for 10 min followed by 2 min at 75 °C and then ligated using Rapid DNA Ligation kit (Roche, Basel, Switzerland), with samples incubated for 5 min at 20 °C. As a negative control, the same reaction was performed without insert addition. One Shot™ TOP10 chemically competent E. Coli cells (Thermo Fisher Scientific, Waltham, MA, USA) were transformed, through heat-shock, with either the ligation product or the negative control, and plated on Agar plates with LB medium and ampicillin. Few resistant colonies were incubated in LB selective medium for 8 h. A backup plate for the selected colonies was stored at 4 °C. Plasmid DNA was extracted with EuroGOLD plasmid Miniprep kit (Euroclone, Pero, MI, Italy) and sequenced (Eurofins Genomics, Vimodrone, MI, Italy) to check proper cloning using the primers in Table 2. Plasmid DNA with the correct integrated insert was amplified starting from the corresponding backup colonies and extracted with NucleoBondXtra Midi Plus kit (Macherey-Nagel, Düren, Germany).

Table 1. PCR primers.

3′UTR *EFEMP1* Forward	5′-AATTGCTAGCTTGACAATAATAGTGGGGCCA-3′
3′UTR *EFEMP1* Reverse	5′-AATTTCTAGATGCCCACTTTATACCATGG-3′

Table 2. Primers for sequencing.

pmirGLO Forward	5′-CGCGAGATTCTCATTAAGGCC-3′
pmirGLO Reverse	5′-CAACTCAGCTTCCTTTCGG-3′

The plasmid DNA containing the cloned *EFEMP1* 3′UTR sequence was used to generate pmiRGLO plasmids carrying a mutated form of the miR-9 target site, using GENEART Site-Directed Mutagenesis System (Thermo Fisher Scientific, Waltham, MA, USA). Specific primers were designed to be used as templates in the mutagenesis reaction (Table 3). Plasmid DNA was extracted from six random grown colonies and sequenced to check for mutated products.

Table 3. Template primers for mutagenesis (mutated sites underlined).

MiR-9 binding site	5′-CCAAAGA-3′
3′UTR *EFEMP1* MUT Forward	5′-ATAAAATAGTGCTTTAAGGTAACAATATCGTGTCGCTGACTTAAA TGCCTGTGGTTGACTCT-3′
3′UTR *EFEMP1* MUT Reverse	5′-AGAGTCAACCACAGGCATTTAAGTCAGCGACACGATATTGTTAC CTTAAAGCACTATTTTAT-3′

2.5. Luciferase Reporter Assay

3×10^5 HEK293 cells were seeded in 12-well plates and co-transfected with 500 ng pmirGLO vector plasmid carrying either the wild-type or the mutated *EFEMP1* 3′UTR and 100 nM miR-9 precursor or negative control, using Lipofectamine 3000 transfection reagent (Thermo Fisher Scientific, Waltham, MA, USA). Cell lysates were collected 24 h post transfection and Firefly and Renilla luciferase activities were quantified by Dual-Luciferase Reporter Assay System (Promega, Madison, WI, USA) on a GLOMAX 20/20 luminometer (Promega, Madison, WI, USA). Firefly luciferase was normalized on Renilla luciferase and the reporter activity was finally expressed as relative activity between cells silenced for miR-9 and the corresponding control.

2.6. Motility Assays

Migration and invasion assays were performed using Transwell Permeable Support 8.0 μm (Corning Incorporated, Corning, NY, USA). 1×10^5 transfected cells in 300 μL of FBS-free medium were seeded in the upper chamber; for invasion, 50 μL of Matrigel (Corning Incorporated, Corning, NY, USA) was added at the bottom of the upper chamber. 10% FBS enriched medium was added to the lower chamber as chemoattractant. After an overnight incubation at 37 °C, migrated/invaded cells were fixed with 100% cold ethanol, stained with 0.4% Sulforhodamine B (GE Healthcare Life Sciences, Chicago, IL, USA) and captured in photos (4 images per well, 10× magnification). For wound-healing assays, 1×10^5 transfected fibroblasts were seeded in 12-well plates. When confluent, cells were removed in the middle of the well with a plastic tip. Images of the wound were captured at this moment and after 48 h (2 images per well, 10× magnification). All images were captured using EVOS XL Core Imaging System (Thermo Fisher Scientific, Waltham, MA, USA) and processed with ImageJ informatic program (NIH, Bethesda, MD, USA).

2.7. Protein Extraction and Western Blot

Whole cell lysates were prepared using NTG buffer (50 mM Tris HCl, 150 mM NaCl, 1% Triton), supplemented with protease inhibitors (Sigma-Aldrich, St. Louis, MO, USA) and activated orthovanadate (1:50). Bradford assay with CoomassiePlus Protein Assay Reagent (Thermo Fisher Scientific, Waltham, MA, USA) was used to quantify the total proteins at Ultrospec 2100 pro (GE Healthcare, Chicago, IL, USA) spectrophotometer. 30 μg total protein were electrophoretically separated on NuPAGE 4–12% Bis-Tris Gel (ThermoFisher Scientific, Waltham, MA, USA). Western blot analyses were performed with primary antibodies: anti-β-actin peroxidase-linked (1:30,000, clone: AC-15, catalog number: A3854, Sigma-Aldrich, St. Louis, Missouri, USA); anti-fibulin-3 (1:200, clone: C-3, catalog number: sc-365224 Santa Cruz Biotechnology, Dallas, TX, USA); anti-e-cadherin (1:200, clone: G-10, catalog number: sc-8426 Santa Cruz Biotechnology, Dallas, TX, USA) and the corresponding secondary antibodies anti-mouse and anti-rabbit peroxidase-linked (1:5000 and 1:10,000, respectively, GE Healthcare, Chicago, IL, USA). The signals were visualized by ECLTM Prime Western Blotting Detection Reagent (GE Healthcare, Chicago, IL, USA). The quantification of protein bands was performed by Quantity One 1-D Analysis (Bio Rad, Hercules, CA, USA).

2.8. Immunohistochemistry

IHC evaluation of fibulin-3 levels was performed on tumor samples collected from the in vivo experiment illustrated in the work by Baroni et al., 2016 (11) (6 samples per experimental condition). Tissue sections were deparaffinised, rehydrated and heated for 5 min at 95 °C in citrate buffer (4:1 sodium citrate (10 mM, pH 8) and citric acid (5 mM); final pH 6). Peroxidase blocking was achieved with 15 min incubation in 80% methanol and 3% hydrogen peroxide. Sections were then incubated with Protein Block Serum-Free (Dako products, Agilent Technologies, Santa Clara, CA, USA) in BSA 1%. Slides were then incubated at room temperature for 1h with a mouse monoclonal anti-fibulin-3 antibody (1:100, clone: C-3, catalog number: sc-365224, Santa Cruz Biotechnology, Dallas, TX, USA) and then with Biotinylated anti-mouse secondary antibody (1:100, Dako) for 45 min. Antibodies were diluted in "Dako real antibody diluition" (Dako products, Agilent Technologies, Santa Clara, CA, USA). Follows HRP-conjugated streptavidin (1:300) for 30 min, DAB (1:50 in HRP substrate buffer) staining for 5 min and mayer's hematoxylin counterstaining for 10 s. Sections were finally dehydrated and mounted. A positivity score ranging from 0 to 2 was assigned to each tumor, having 0 for no signal, 1 for intermediate positivity and 2 for high positivity.

2.9. Tumor Cell Conditioning and Resistance Test

On the first day, 4.5×10^5 immortalized fibroblasts were seeded in 6-well plates. After 24 h, NFs were transfected with either miR-9 or si-*EFEMP1* and controls, and 3×10^5 MDA-MB-468 cells were seeded in 6-wells plates. On the third day, MDA-MB-468 cells were conditioned with the medium of transfected NFs and then treated (or not) with Cisplatin (5 µM) after 24 h. The drug was added in fresh medium. On day 5, cell viability was assessed by cell counting.

2.10. Mining Data to Evaluate Correlation of MiR-9 Expression and Cisplatin Response

Publicly available data from TNBC data sets with available matched mRNA-miRNA expression profiles from The Cancer Genome Atlas (GDC TCGA Breast Cancer RNA counts) were downloaded from the Xena browser, while normalized data from METABRIC study [17] were recovered through cBiportal, together with our in house cohort (SubSeries GSE86948). Genes from each platform were annotated with biomaRt and only common cross-platform genes were selected for further analysis. TCGA data were downloaded as raw counts and processed with limma-voom in limma R package. Normalized data were scaled by median-absolute-deviation (MAD) for each sample. For TCGA miRNA expression profiles, TPM data was downloaded from TCGA BRCA cohort in XENA.

Gene expression signatures were explored for their correlation with the CAF populations identified by dedicated metagenes reported by Bartoschek M et al. [18]. The included endothelial/microvasculature signature [19], stroma-related signature [20] and microvasculature signature [21]. Gene signature scores were computed as the averages of mean centred expression of all these gene members of each signature. For each metagene, correlation patterns were compacted using Pearson correlation.

3. Results

3.1. In-Silico Evaluation of EFEMP1 Levels in CAFs

Aiming at investigating *EFEMP1* role in the conversion of normal to cancer-associated fibroblasts in the breast cancer microenvironment, we analyzed its expression level in six matched paired NFs/CAFs obtained from breast malignances (two grade III, three grade II and one grade I; GSE20086). Figure 1a illustrates the significant downregulation of *EFEMP1* in CAFs vs. their matched NFs.

Moreover, since breast cancer is a complex and highly heterogeneous disease, to gain a better understanding of these complexities we analyzed *EFEMP1* expression in public profiles of human dermal fibroblasts conditioned with three breast cancer cell line models (GSE80035). Relevantly, fibroblasts conditioned with TNBC (MDA-MB-468) and HER2+ (SkBr3) cells presented a lower *EFEMP1* expression than Luminal A ER+/PR+/HER2+ (T-47D) cells (Figure 1b). In support of these observations, CAFs isolated from human TNBC tumors (GSE37614) presented a lower expression of *EFEMP1* in comparison to other tumor subtypes (Figure 1c). These data are strengthened by the result of qRT-PCR analysis of *EFEMP1* expression in a couple of NFs/CAFs from a TNBC patient, illustrated in Figure S1.

Thus, these results suggest that *EFEMP1* downmodulation is linked to the acquisition of a malignant phenotype in tumor-associated fibroblasts, which seems to be particularly relevant in TNBC subtype.

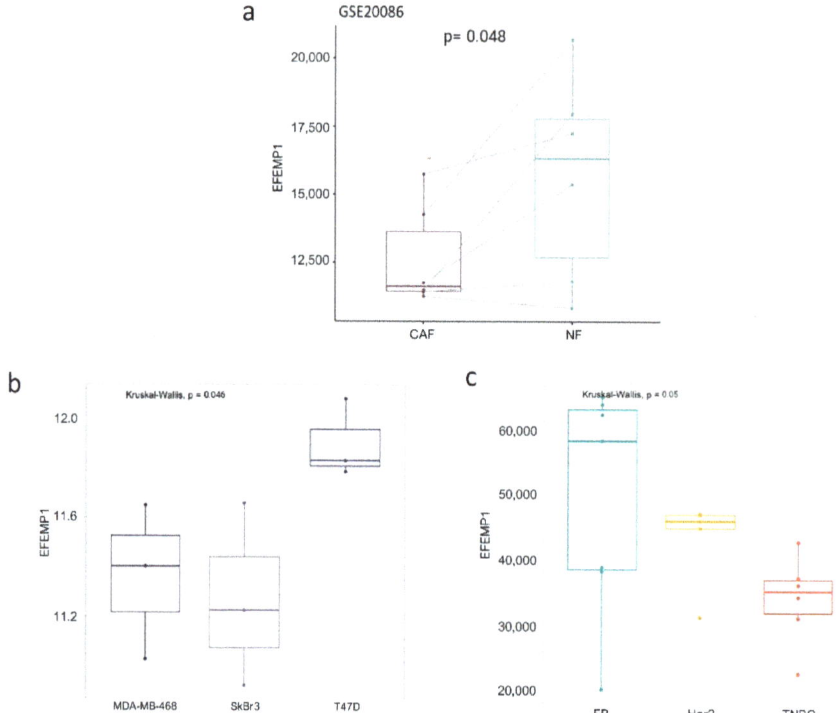

Figure 1. *EFEMP1* is downregulated in breast cancer-associated and TNBC-conditioned fibroblasts. In-silico evaluation of *EFEMP1* levels in paired NFs/CAFs of six breast cancer patients (**a**); in normal human dermal fibroblasts conditioned with the supernatants of breast cancer cells of different subtypes (**b**) and early passage of primary CAFs isolated from human breast cancer samples classified as ER+ (*n* = 7), TNBC (*n* = 7) and HER2+ (*n* = 6) (**c**).

3.2. MiR-9 Directly Targets EFEMP1 and Affects Protein Levels In Vitro and In Vivo

Encouraged by the in-silico results, we proceeded assessing *EFEMP1* expression in our normal fibroblast in vitro model (NFs) at mRNA and protein level, upon miR-9 transfection, by qRT-PCR and western blot analyses, respectively. As shown in Figure 2a,b, *EFEMP1* and fibulin-3 levels decreased in miR-9 overexpressing NFs (NFs miR-9) compared to control (NFs miR-NEG).

Fibulin-3 is a secreted protein and it exerts its main activity as anchoring element in the stroma. In order to verify miR-9-induced *EFEMP1* downmodulation in this cellular compartment, we performed an IHC analysis on tumor samples from our previous in vivo experiment. Particularly, it was monitored the in vivo tumor growth of MDA-MB-468 cells co-injected in the mammary fat pad of SCID mice with NFs transfected with miR-9 (NFs/miR-9) or negative control (NFs/miR-neg), which resulted increased in MDA-MB-468 cells and NFs/miR-9 group [11]. Thus, evaluating fibulin-3 expression in tumor samples from mice injected with MDA-MB-468 and NFs/ miR-9 compared to negative control, we observed a lower expression of this protein in the tumor stroma (Figure 2c and Figure S2a). Since MDA-MB-468 and NFs/miR-9 mice developed bigger tumors compared to negative control, it is reasonable to hypothesize an anti-oncogenic role for this ECM protein in the TNBC stroma.

Even though a slight decrease in fibulin-3 levels was observed also in some of the tumor nodules in the MDA-MB-468 + NFs miR-9 group, no modulation of *EFEMP1*/fibulin-3 expression was detected in MDA-MB-468 cells overexpressing miR-9 in in vitro experiments (Figure S2b). We evaluated e-cadherin as positive control since it has been already validated as miR-9 target in tumor cells. Thus, these results suggest that *EFEMP1* is not a miR-9 target in this cell model.

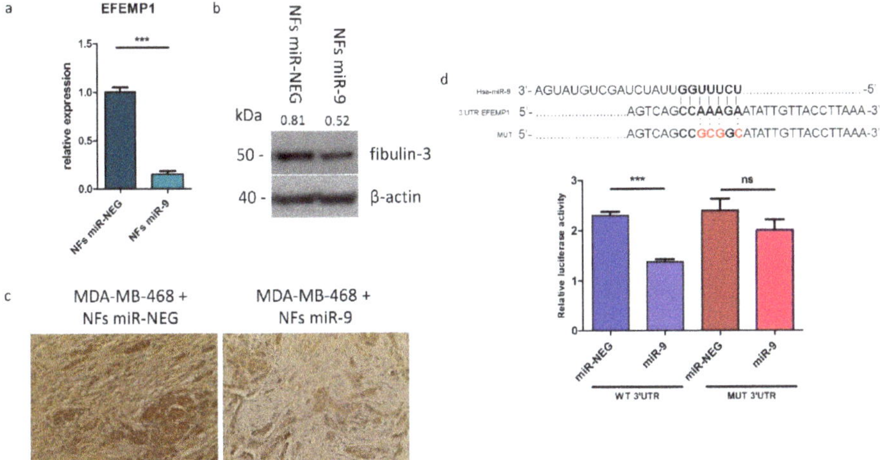

Figure 2. *EFEMP1* is a direct target of miR-9. Evaluation of *EFEMP1* gene and protein levels by qRT-PCR (**a**), western blot (**b**) and IHC (**c**). qRT-PCR and western blot analysis were performed on NFs miR-9 vs. control. Protein expression levels are indicated above western blot bands. IHC images show fibulin-3 expression in ex vivo samples of tumors grown from the co-injection of MDA-MB-468 cells and NFs miR-NEG/9. Images are representative; the experiment was performed on 6 tumors per group. Scale bars 2.5 μm (**d**). Luciferase assay performed on HEK293 cell line transfected with miR-9 or control and with wild-type or mutated *EFEMP1* 3′UTR (mutated sequence shown above). Data are presented as the mean of three biological replicates ±SEM (*** $p < 0.001$, ns = non-significant).

In order to check whether *EFEMP1* regulation by miR-9 in fibroblasts is due to a direct targeting, we performed a luciferase reporter assay. Wild-type or mutated *EFEMP1* 3′UTR were cloned downstream the luciferase gene and co-transfected with miR-9 or control in HEK-293T cells. As illustrated in Figure 2d, we observed a significant reduction of the luciferase activity in the cells transfected with the wild-type construct in the presence of miR-9, compared to control. This effect was lost when the mutated 3′UTR was tested.

3.3. EFEMP1 Silencing Recapitulates miR-9-Induced CAF-Like Features in Normal Fibroblasts

To evaluate the contribution of *EFEMP1* downmodulation to the acquisition of CAF-like features upon miR-9 targeting, we first performed migration and invasion assays. Normal fibroblasts were transfected with siRNA targeting *EFEMP1* (si-*EFEMP1*) or with a negative control (si-NEG). As shown in Figure 3a,b, *EFEMP1* knockdown significantly increased fibroblast motility. Specifically, at 24 h, a +15% of cells migrated to the bottom chamber of the transwell, while +28% of cells invaded the Matrigel upon *EFEMP1* silencing, compared to control. In order to better appreciate si-*EFEMP1* phenocopy of miR-9 effect, we decided to perform a wound healing assay on fibroblasts transfected in parallel with miR-9 or si-*EFEMP1* vs. each respective control. Figure 3c shows that both miR-9 overexpression and *EFEMP1* silencing increased fibroblasts ability to "heal the wound", evaluated 48 h after the scratch. For each experiment, transfection efficiency was assessed by qRT-PCR (Figure S3). Thus, we demonstrated that *EFEMP1* silencing partially mimics miR-9 action in NFs, leading to the acquisition of CAF-like features.

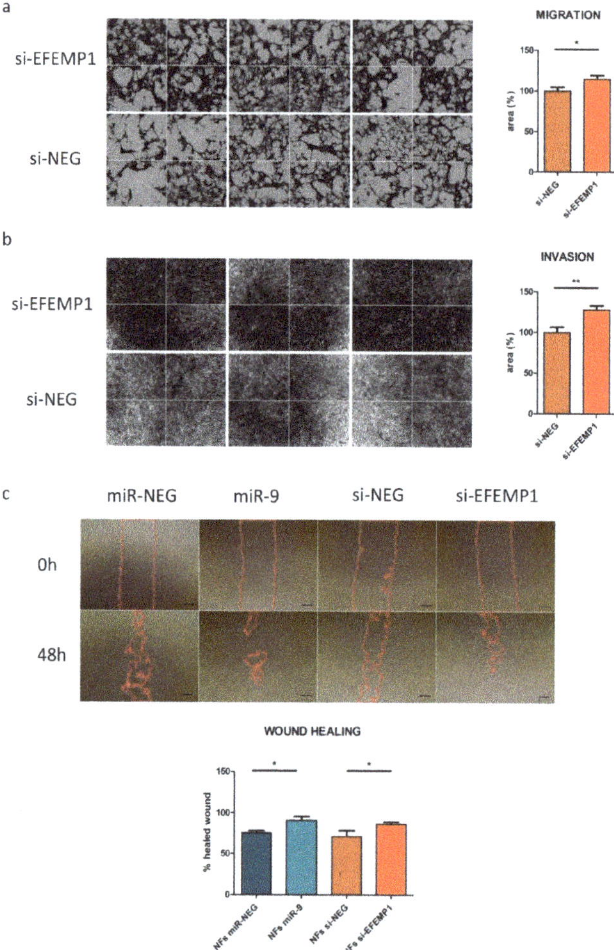

Figure 3. *EFEMP1* silencing increases fibroblast's motility. Migration (**a**), invasion (**b**) and wound healing (**c**) assays performed on fibroblasts transfected with miR-9 (wound healing exclusively)/si-*EFEMP1* or controls. In Figure 3c, the red line identifies the region of the wound which is still not occupied by cells. Images are representative and data are presented as mean of three biological replicates ±SEM. (* $p < 0.05$; ** $p < 0.01$); scale bars, 100 μm.

3.4. CAF-Like Properties Induced by miR-9/si-EFEMP1-Transfection Reduce MDA-MB-468 Cell Sensitivity to Cisplatin

It is well known that CAFs can also affect tumor cell responsiveness to treatment by triggering multiple escape mechanisms. For instance, Figure S4 shows *EFEMP1* mRNA pattern among CAFs isolated from tumors of sensitive and resistant breast cancer patients before neo-adjuvant chemotherapy. CAFs from resistant patients exhibited slightly lower *EFEMP1* mRNA levels than sensitives. Since platinum-based therapy is an effective treatment for a subset of TNBCs [22], we then decided to evaluate the ability of miR-9/si-*EFEMP1*-induced CAF-like cells to affect tumor cell sensitivity to the anti-cancer drug cisplatin. MDA-MB-468 cells were chosen among the available TNBC cell lines considering their sensitivity to this compound [23] and our existing expertise with this cell model.

Tumor cells were conditioned for 24 h with the supernatant of NFs miR-9/si-*EFEMP1* or controls, and then treated with cisplatin (5 μM, IC50 concentration) for 24h. When we challenged the tumor cells with cisplatin, we observed a 15% increase in MDA-MB-468 cell viability upon conditioning with NFs miR-9 supernatant, compared with control conditions (Figure 4a,b). Transfection efficiencies related to this experiment are shown in Figure S5a.

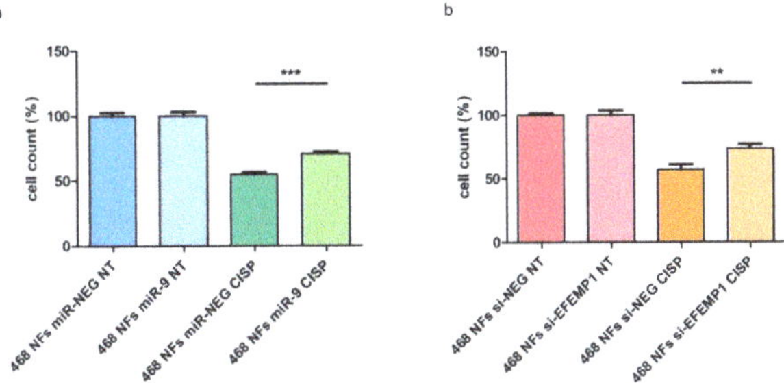

Figure 4. NFs miR-9/siEFEMP1 reduce tumor cell sensitivity to cisplatin. MDA-MB-468 cell count upon treatment with cisplatin (24 h) after 24h of conditioning with the supernatant of (**a**) NFs miR-9 or (**b**) si-EFEMP1, compared to controls. Cell count data are presented as mean of the percentage of viable treated (CISP = cisplatin) cells of three biological replicates, compared to non-treated (NT) cells, ±SEM (** $p < 0.01$, *** $p < 0.001$).

It is worth noting that we detected an increase in miR-9 levels in MDA-MB-468 cells conditioned with the supernatant of NFs miR-9 (Figure S5b). This could be due to miR-9 uptake by MDA-MB-468 cells from NF medium. However, a slight but significant miR-9 upmodulation was also seen in treated control cells, compared to the non-treated counterpart, suggesting an additional action of the treatment alone on tumor miR-9 levels. Further studies should be performed to investigate the biological meaning of these data.

This evidence demonstrates the relevance of miR-9/*EFEMP1* axis on the transition of NFs phenotype to CAF-like, which, in turn, promotes chemoresistance in TNBC.

3.5. Characterization of miR-9/CAF Axis on TNBC Biology and Chemotherapy Response by Mining mRNA and miRNA Expression Data

To further analyze whether miR-9/CAF axis on TNBC is related with cisplatin treatment response we analyzed the transcriptional landscape on TNBC and public available signatures. Recently, single-cell resolution analysis revealed the existence of at least two spatially and functionally subsets of breast CAFs: (1) vascular CAFs (vCAFs), enriched in vascular development and angiogenesis signaling pathways and (2) matrix CAFs (mCAF), endowed in matrix-related genes and stroma-related treatment-predictive signatures [18].

To further identify functionally distinctive CAFs through reported molecular signatures we analyzed the transcriptional landscape of TNBC from the public data sets TCGA and METABRIC, as well as an in-house profiled cohort (GSE86948) composed of mRNA-miRNA matched expression profiles (*n* = 342). Notably, on TCGA and GSE86948 datasets, a similar expression pattern of miR-9 was observed in matched normal adjacent tissue and tumor cells of TNBC patients (Figure S6a,b), suggesting a coordinate and correlated altered phenotype in both breast tissues (Figure S6c). Consequently, the tumoral miR-9 expression pattern is informative of the miRNA expression in the stroma comportment.

We then sub-grouped TNBC data sets according to miR-9 level as following: miR-9 high (over 3rd Quantile), intermediate (Inter, >3rd Q and <1st Q) and low (<1st Q). We first set out to determine whether the observed CAF subtypes, detected by dedicated metagenes, are correlated with their inferred functions, including modulation of extracellular matrix production (ECM metagene) and angiogenesis (endothelial metagene) (Table S1) [19]. In keeping with reported data, the vCAF signature was highly correlated to an endothelial cell metagene ($R = 0.61$, $p < 0.01$ vs $R = 0.28$ in mCAF) and microvascular signature ($R = 0.61$, $p < 0.01$ vs $R = 0.3$ in mCAF) (Figure S7a), whereas the mCAF signature was strongly associated with the ECM metagene ($R = 0.98$, $p < 0.01$ vs 0.49 in vCAF) and stroma signature ($R = 0.98$, $p < 0.01$ vs 0.55 in vCAF) (Figure S7b).

Furthermore, correlations within TNBC tumors were dependent on the miR-9 subgroup. Notably, the relations between CAFs and gene signatures in tumors with high or intermediate miR-9 expression strongly indicate that the functionality of both ECM and endothelial gene programs correlated with vCAFs and mCAFs. In contrast, tumors with low miR-9 expression present a dependent relation of endothelial signature only in vCAF (Figure 5a and Figure S7a,b). These specific correlated profiles further indicate the existence of different CAF subtypes in TNBC related with elevated miR-9 expression, and represent a strong support of the notion that miR-9 up-modulation modifies NFs, which in turn support malignant phenotypes and likely provide advantages against chemotherapy treatment.

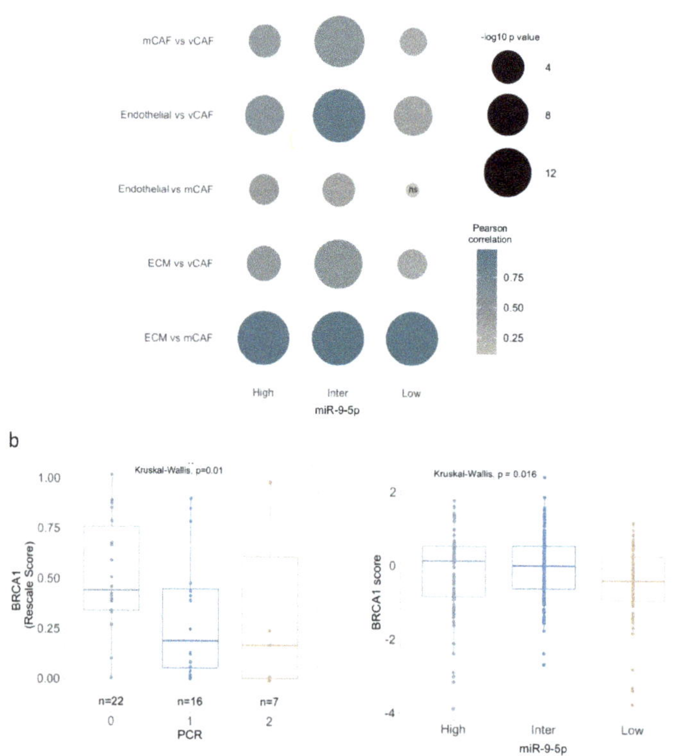

Figure 5. miR-9 expression in TNBC correlates with different CAF subsets and resistance to cisplatin. (**a**) Bubble plot showing computed Pearson correlations between TNBC subgroups according to miR-9 expression, CAF subsets and biological signatures of related functions. ns: non-significant, significant p value < 0.05. Bubble colour represents Person correlation, while size corresponds to −log10 p value, as illustrated in plot legend. (**b**) BRCA1 gene expression in two well-characterized cohorts of patients with TNBC treated in neoadjuvance with cisplatin (GSE18864 and GSE103668), evaluated Pathological complete response by Miller-Payne (MP) criteria (0: MP 0, 1, 2 Progression, no change or still high tumor cellularity; 1: MP 3, 4 minor and marked loss of tumor cells, 2: MP 5 non-malignant cells) (left panel), and BRCA1 gene expression in TNBC sub-grouped by miR-9 expression (right panel).

We therefore investigated whether miR-9 conveys sensitivity to therapy in human TNBC tumors. Relevantly, literature has reported that low BRCA1 mRNA expression is a factor associated with good cisplatin response [24]. Thus, we examined BRCA1 gene expression in two well-characterized cohorts of patients with TNBC treated in neoadjuvant with cisplatin (GSE18864 and GSE103668). Patients with lower BRCA1 expression respond better to cisplatin treatment, compared to patients expressing moderate or high BRCA1 levels, evaluated by Miller–Payne criteria (Figure 5b). This is consistent with the idea that "BRCAness" phenotype is characterized by a decreased BRCA1 expression [25,26]. Relevantly, a similar BRCA1 expression pattern was observed in TNBC tumors sub-grouped by miR-9 expression; for instance, high and intermediate miR-9 category displayed a significantly higher BRCA1 expression. Together, these data provide independent evidence that miRNA signaling, other than prompting a fibroblast reprogramming, can also affect response to cisplatin, likely by modulating CAF/tumor interplay.

4. Discussion

Given the idea of a tumor tissue as "a wound that never heals", the tumor microenvironment can also be chronically altered through a reciprocal tumor–stroma signaling. Indeed, CAFs, which constitute the major component in the stroma, exert several pro-tumoral functions [27]. It is generally accepted that CAFs, considered fibrotic myofibroblasts, have distinctive features, functions or location from normal fibroblasts, and contribute to establish and maintain the aggressiveness of the lesion.

Approximately 80% of fibroblasts in breast cancer stroma acquires an aggressive phenotype [28]; however, how such activation occurs is still not well understood. In our previous work, we unravelled one of the mechanisms engaged by TNBC cells to obtain fibroblast's support. We provided evidence that TNBC cells overexpressing miR-9 are able to release the miRNA into the stroma, where normal fibroblasts are able to incorporate it. Consequently, miR-9 perturbs the transcriptional landscape of the recipient cells, inducing a shift towards CAF malignant phenotype. The data presented here extended these findings and demonstrated that *EFEMP1* downregulation, due to direct miR-9 regulation, is a relevant step in the malignant transformation of fibroblasts in the TNBC microenvironment. We also showed that *EFEMP1* specific silencing in NFs partially recapitulates the CAF-like features triggered by miR-9 uptake, such as an increased ability to migrate and invade. Certainly, considering the common mechanism of action of microRNAs, able to finely tune several molecules to achieve a specific biological effect, it is conceivable that miR-9 has additional targets implicated in fibroblast's behaviour, and it would be interesting to explore other candidates.

Another important oncogenic downstream effect of CAF reprogramming includes the impairment of chemotherapy efficacy. The mechanisms underlying this process still have to be fully elucidated, but the literature already provides interesting inputs. For example, CAFs can convey pro-survival cues to tumor cells, induce epithelial-to-mesenchymal transition, angiogenesis, metabolic reprogramming and stemness traits [29–31]. Interestingly, in a dataset comparing gene expression of CAF from breast cancer patients resistant *vs* sensitive to neoadjuvant chemotherapy, *EFEMP1* was found significantly downmodulated in the resistant group. Moreover, this CAF subgroup was associated to cancer stemness phenotype, a feature associated to disease aggressiveness and resistance to chemotherapy [32]. It is interesting to note that Bartoschek and collaborators reported that the absolute number of CAFs in tumor tissues before receiving neoadjuvant chemotherapy is not statistically different between sensitive and resistant patients [19]; instead the CAF subclasses defined in their study and also analysed in the present work are differentially operating in each tumor class and, relevantly, presented a distinctive correlation with miR-9 expression. In particular, correlation data of miR-9 overexpressing tumors (high an intermediate subgroups) and CAFs subsets pinpoint the functional differences driven by miR-9/ CAF axis. Indeed, in miR-9 overexpressing tumors ECM and endothelial gene programs correlate with both vCAFs and mCAFs, tumors with low miR-9 expression present a dependent relation of endothelial signature only in vCAF. Interestingly, mCAFs are highly associated with a stroma-derived invasion signature predictive of responsiveness to neoadjuvant chemotherapy in breast cancer [18]. Numerous clinical trials are currently revaluating cisplatin as chemotherapeutic option to treat TNBC, especially those harbouring a BRCA mutation [24,33]. As expected, our data show that lower BRCA1 expression is found in cisplatin responder patients, compared to non-responders. BRCA1 expression analysis in TNBC tumors, sub-grouped on the basis of miR-9 expression, revealed that tumors with higher miR-9 expression in tumor or fibroblast compartment also over-expressed BRCA1, further supporting the correlation of high miR-9 expression to a chemo-resistance phenotype. Consistently, our in vitro experiments corroborate this hypothesis: miR-9/si-*EFEMP1*-induced CAF-like cells were able to impact on TNBC cell sensitivity to cisplatin. MDA-MB-468 cells conditioned with the supernatant of either miR-9 or si-*EFEMP1* transfected NFs resulted in a significant increment of viable cells after treatment, compared to control. Moreover, our data show a moderate increase of miR-9 levels in MDA-MB-468 cells conditioned with the supernatant of NFs transfected with miR-9. This event could be the result of either an uptake or/and an induction of the miRNA upon cell conditioning and contributes to the observed resistant phenotype. Indeed, a recent review reports a list of CAF-secreted miRNAs responsible of conferring cisplatin resistance in different tumor models, even though miR-9 was not reported [34]. However, the reduction of sensitivity in conditioned tumor cells can also be caused by other multiple secreted factors rather than by a single molecule. Further studies should be performed to explore these mechanisms. Considering that TNBC patients still lack targeted therapies and rely only on standard chemotherapy, our data appear particularly relevant for future translational studies. The literature extensively suggests the perspective of depleting CAFs to ameliorate patient's prognosis, but no relevant results were obtained so far [35]. Another proposed approach is the CAF reversion to a non-malignant phenotype. Since miR-9 was demonstrated to act on multiple targets, both in breast cancer and stromal cells, it would be advantageous to exploit this target for therapeutic purposes [36–38].

Certainly, since one of the main concerns about a miRNA-derived therapy is the potential side effects, especially when considering a miRNA that seems to have contrasting roles in different tumor types and/or tissues, the most successful approach to overcome this issue would likely be to develop a tumor-specific delivery [39]. Data in the literature regarding *EFEMP1* expression in the tumor epithelium are also controversial: it was found downregulated in lung, nasopharyngeal, prostate, hepatocellular and glioma cancers, compared to normal tissue [40–43]; on the contrary, it acts as oncogene in cervical, pancreatic and ovarian cancers [44–46]. In breast cancer, *EFEMP1* was found downmodulated in sporadic malignancies but there is also evidence of pro-tumor activities [47,48]. Moreover, our qRT-PCR and western blot evaluation of *EFEMP1* levels in miR-9 transfected MDA-MB-468 cells suggests that the fibulin is not a target of the miRNA in this cell line model.

Furthermore, it is important to note that fibroblasts are the main secreting cell compartment of fibulin-3 in the stroma. The molecule exerts its principal activity as structural protein, although it is also known to induce and interact with the tissue inhibitor of metalloproteinase-3 TIMP-3, which inhibits metalloproteinases MMP2/9, highly expressed in breast cancers and actively involved in matrix remodelling. IHC evaluation of fibulin-3 levels in ex vivo samples suggests that a reduced expression of the protein in the stroma milieu could have provided an oncogenic advantage to MDA-MB-468 + NFs miR-9 tumors, given that this group grew significantly more than controls [11].

In conclusion, our results demonstrate that miR-9 directly targets *EFEMP1* in fibroblasts and that *EFEMP1* downmodulation is important in determining NF's acquisition of CAF-like properties. Additional experiments are necessary to address the intriguing fibroblast-specific miR-9 targeting of *EFEMP1* in tumoral cells. Our work sheds light on previously unknown mechanisms that define NFs reprogramming in TNBC and has significant therapeutic implications for patients with this tumor subtype.

Supplementary Materials: The following are available online at http://www.mdpi.com/2073-4409/9/9/2143/s1, Figure S1. *EFEMP1* expression in TNBC paired NFs/CAFs. Figure S2. Evaluation of fibulin-3 levels in vivo and in vitro upon miR-9 overexpression in NFs. Figure S3. miR-9/si-*EFEMP1* transfection efficiency of fibroblasts used in motility assays. Figure S4. In-silico evaluation of *EFEMP1* expression in CAFs isolated from resistant vs sensitive breast cancer patients treated with neoadjuvant chemotherapy. Figure S5. Evaluation of fibroblasts transfection efficiency and analysis of miR-9 levels in conditioned and treated MDA-MB-468. Figure S6. In-silico evaluation of miR-9 expression in matched adjacent normal and tumor tissue of TNBC patients from TCGA and GSE38167 data sets among breast cancer subtypes (HER+, HR+/HER2+ and TN). Figure S7. Correlation analysis between vCAF (a) and mCAF (b) subsets and ECM, endothelial, microvasculature and stroma gene signatures. Table S1. Publicly available metagenes used to detect respectively vCAF, mCAF, endothelial and ECM CAF subtypes.

Author Contributions: G.C. and M.V.I. conceived the project; G.C., A.C. wrote the manuscript. G.C. and I.P. performed the experiments and analyzed the data. S.R.-C. performed the bioinformatic analysis, wrote paper sections, discussed results and revised the manuscript. M.V.I. supervised the project. M.V.I. and A.C. revised the manuscript. All authors have read and agreed to the published version of the manuscript.

Funding: This research was funded by Berlucchi Career Foundation grant and Young Investigator grant from Italian Ministry of Health (GR-2016-02361750) to M.V. Iorio. Alessandra Cataldo was supported by Fondazione Umberto Veronesi fellowship.

Acknowledgments: Thanks to Cristina Ghirelli for help with cell culture. Primary NFs and CAFs were kindly provided by Cinzia De Marco (Department of Applied Research and Technical development). Thanks to Giovanni Centonze (Department of Pathology and Laboratory Medicine) for helping in the quantification of IHC images.

References

1. Huet, E.; Jaroz, C.; Nguyen, H.Q.; Belkacemi, Y.; de la Taille, A.; Stavrinides, V.; Whitaker, H. Stroma in normal and cancer wound healing. *FEBS J.* **2019**, *286*, 2909–2920. [CrossRef] [PubMed]

2. Quail, D.F.; Joyce, J.A. Microenvironmental regulation of tumor progression and metastasis. *Nat. Med.* **2013**, *19*, 1423–1437. [CrossRef] [PubMed]

3. Dumont, N.; Liu, B.; Defilippis, R.A.; Chang, H.; Rabban, J.T.; Karnezis, A.N.; Tjoe, J.A.; Marx, J.; Parvin, B.; Tlsty, T.D. Breast fibroblasts modulate early dissemination, tumorigenesis, and metastasis through alteration of extracellular matrix characteristics. *Neoplasia* **2013**, *15*, 249–262. [CrossRef] [PubMed]

4. Yu, Y.; Xiao, C.-H.; Tan, L.-D.; Wang, Q.-S.; Li, X.-Q.; Feng, Y.-M. Cancer-associated fibroblasts induce epithelial-mesenchymal transition of breast cancer cells through paracrine TGF-β signalling. *Br. J. Cancer* **2014**, *110*, 724–732. [CrossRef] [PubMed]

5. Fiori, M.E.; Di Franco, S.; Villanova, L.; Bianca, P.; Stassi, G.; De Maria, R. Cancer-associated fibroblasts as abettors of tumor progression at the crossroads of EMT and therapy resistance. *Mol. Cancer* **2019**, *18*, 70. [CrossRef]

6. Liao, D.; Luo, Y.; Markowitz, D.; Xiang, R.; Reisfeld, R.A. Cancer associated fibroblasts promote tumor growth and metastasis by modulating the tumor immune microenvironment in a 4T1 murine breast cancer model. *PLoS ONE* **2009**, *4*, e7965. [CrossRef]

7. Hembruff, S.L.; Jokar, I.; Yang, L.; Cheng, N. Loss of transforming growth factor-beta signaling in mammary fibroblasts enhances CCL2 secretion to promote mammary tumor progression through macrophage-dependent and -independent mechanisms. *Neoplasia* **2010**, *12*, 425–433. [CrossRef]

8. Orimo, A.; Gupta, P.B.; Sgroi, D.C.; Arenzana-Seisdedos, F.; Delaunay, T.; Naeem, R.; Carey, V.J.; Richardson, A.L.; Weinberg, R.A. Stromal fibroblasts present in invasive human breast carcinomas promote tumor growth and angiogenesis through elevated SDF-1/CXCL12 secretion. *Cell* **2005**, *121*, 335–348. [CrossRef]

9. Yu, T.; Di, G. Role of tumor microenvironment in triple-negative breast cancer and its prognostic significance. *Chin. J. Cancer Res.* **2017**, *29*, 237–252. [CrossRef]

10. Kogure, A.; Kosaka, N.; Ochiya, T. Cross-talk between cancer cells and their neighbors via miRNA in extracellular vesicles: An emerging player in cancer metastasis. *J. Biomed. Sci.* **2019**, *26*, 7. [CrossRef]

11. Baroni, S.; Romero-Cordoba, S.; Plantamura, I.; Dugo, M.; D'Ippolito, E.; Cataldo, A.; Cosentino, G.; Angeloni, V.; Rossini, A.; Daidone, M.G.; et al. Exosome-mediated delivery of miR-9 induces cancer-associated fibroblast-like properties in human breast fibroblasts. *Cell Death Dis* **2016**, *7*, e2312. [CrossRef] [PubMed]

12. Li, X.; Zeng, Z.; Wang, J.; Wu, Y.; Chen, W.; Zheng, L.; Xi, T.; Wang, A.; Lu, Y. MicroRNA-9 and breast cancer. *Biomed. Pharmacother.* **2020**, *122*, 109687. [CrossRef] [PubMed]

13. De Vega, S.; Iwamoto, T.; Yamada, Y. Fibulins: Multiple roles in matrix structures and tissue functions. *Cell. Mol. Life Sci.* **2009**, *66*, 1890–1902. [CrossRef] [PubMed]

14. Zhang, Y.; Marmorstein, L.Y. Focus on molecules: Fibulin-3 (EFEMP1). *Exp. Eye Res.* **2010**, *90*, 374–375. [CrossRef] [PubMed]

15. Tian, H.; Liu, J.; Chen, J.; Gatza, M.L.; Blobe, G.C. Fibulin-3 is a novel TGF-β pathway inhibitor in the breast cancer microenvironment. *Oncogene* **2015**, *34*, 5635–5647. [CrossRef] [PubMed]

16. Durinck, S.; Spellman, P.T.; Birney, E.; Huber, W. Mapping Identifiers for the Integration of Genomic Datasets with the R/Bioconductor package biomaRt. *Nat. Protoc.* **2009**, *4*, 1184–1191. [CrossRef]

17. Pereira, B.; Chin, S.-F.; Rueda, O.M.; Vollan, H.-K.M.; Provenzano, E.; Bardwell, H.A.; Pugh, M.; Jones, L.; Russell, R.; Sammut, S.-J.; et al. The somatic mutation profiles of 2,433 breast cancers refines their genomic and transcriptomic landscapes. *Nat. Commun.* **2016**, *7*, 11479. [CrossRef]

18. Bartoschek, M.; Oskolkov, N.; Bocci, M.; Lövrot, J.; Larsson, C.; Sommarin, M.; Madsen, C.D.; Lindgren, D.; Pekar, G.; Karlsson, G.; et al. Spatially and functionally distinct subclasses of breast cancer-associated fibroblasts revealed by single cell RNA sequencing. *Nat. Commun* **2018**, *9*, 5150. [CrossRef]

19. Winslow, S.; Lindquist, K.E.; Edsjö, A.; Larsson, C. The expression pattern of matrix-producing tumor stroma is of prognostic importance in breast cancer. *BMC Cancer* **2016**, *16*, 841. [CrossRef]

20. Farmer, P.; Bonnefoi, H.; Anderle, P.; Cameron, D.; Wirapati, P.; Wirapati, P.; Becette, V.; André, S.; Piccart, M.; Campone, M.; et al. A stroma-related gene signature predicts resistance to neoadjuvant chemotherapy in breast cancer. *Nat. Med.* **2009**, *15*, 68–74. [CrossRef]

21. Tobin, N.P.; Wennmalm, K.; Lindström, L.S.; Foukakis, T.; He, L.; Genové, G.; Östman, A.; Landberg, G.; Betsholtz, C.; Bergh, J. An Endothelial Gene Signature Score Predicts Poor Outcome in Patients with Endocrine-Treated, Low Genomic Grade Breast Tumors. *Clin. Cancer Res.* **2016**, *22*, 2417–2426. [CrossRef] [PubMed]

22. Birkbak, N.J.; Li, Y.; Pathania, S.; Greene-Colozzi, A.; Dreze, M.; Bowman-Colin, C.; Sztupinszki, Z.; Krzystanek, M.; Diossy, M.; Tung, N.; et al. Overexpression of BLM promotes DNA damage and increased sensitivity to platinum salts in triple-negative breast and serous ovarian cancers. *Ann. Oncol.* **2018**, *29*, 903–909. [CrossRef] [PubMed]

23. Lehmann, B.D.; Bauer, J.A.; Chen, X.; Sanders, M.E.; Chakravarthy, A.B.; Shyr, Y.; Pietenpol, J.A. Identification of human triple-negative breast cancer subtypes and preclinical models for selection of targeted therapies. *J. Clin. Investig.* **2011**, *121*, 2750–2767. [CrossRef] [PubMed]

24. Silver, D.P.; Richardson, A.L.; Eklund, A.C.; Wang, Z.C.; Szallasi, Z.; Li, Q.; Juul, N.; Leong, C.-O.; Calogrias, D.; Buraimoh, A.; et al. Efficacy of neoadjuvant Cisplatin in triple-negative breast cancer. *J. Clin. Oncol.* **2010**, *28*, 1145–1153. [CrossRef]

25. Turner, N.C.; Reis-Filho, J.S.; Russell, A.M.; Springall, R.J.; Ryder, K.; Steele, D.; Savage, K.; Gillett, C.E.; Schmitt, F.C.; Ashworth, A.; et al. BRCA1 dysfunction in sporadic basal-like breast cancer. *Oncogene* **2007**, *26*, 2126–2132. [CrossRef]

26. Turner, N.; Tutt, A.; Ashworth, A. Hallmarks of "BRCAness" in sporadic cancers. *Nat. Rev. Cancer* **2004**, *4*, 814–819. [CrossRef]

27. Dvorak, H.F. Tumors: Wounds that do not heal. Similarities between tumor stroma generation and wound healing. *N. Engl. J. Med.* **1986**, *315*, 1650–1659. [CrossRef]

28. Gascard, P.; Tlsty, T.D. Carcinoma-associated fibroblasts: Orchestrating the composition of malignancy. *Genes Dev.* **2016**, *30*, 1002–1019. [CrossRef]

29. Long, X.; Xiong, W.; Zeng, X.; Qi, L.; Cai, Y.; Mo, M.; Jiang, H.; Zhu, B.; Chen, Z.; Li, Y. Cancer-associated fibroblasts promote cisplatin resistance in bladder cancer cells by increasing IGF-1/ERβ/Bcl-2 signalling. *Cell Death Dis.* **2019**, *10*, 375. [CrossRef]

30. Wang, L.; Li, X.; Ren, Y.; Geng, H.; Zhang, Q.; Cao, L.; Meng, Z.; Wu, X.; Xu, M.; Xu, K. Cancer-associated fibroblasts contribute to cisplatin resistance by modulating ANXA3 in lung cancer cells. *Cancer Sci.* **2019**, *110*, 1609–1620. [CrossRef]

31. Kadel, D.; Zhang, Y.; Sun, H.-R.; Zhao, Y.; Dong, Q.-Z.; Qin, L.-X. Current perspectives of cancer-associated fibroblast in therapeutic resistance: Potential mechanism and future strategy. *Cell Biol. Toxicol.* **2019**, *35*, 407–421. [CrossRef] [PubMed]

32. Su, S.; Chen, J.; Yao, H.; Liu, J.; Yu, S.; Lao, L.; Wang, M.; Luo, M.; Xing, Y.; Chen, F.; et al. CD10+GPR77+ Cancer-Associated Fibroblasts Promote Cancer Formation and Chemoresistance by Sustaining Cancer Stemness. *Cell* **2018**, *172*, 841–856.e16. [CrossRef] [PubMed]

33. Hill, D.P.; Harper, A.; Malcolm, J.; McAndrews, M.S.; Mockus, S.M.; Patterson, S.E.; Reynolds, T.; Baker, E.J.; Bult, C.J.; Chesler, E.J.; et al. Cisplatin-resistant triple-negative breast cancer subtypes: Multiple mechanisms of resistance. *BMC Cancer* **2019**, *19*, 1039. [CrossRef] [PubMed]

34. Santos, P.; Almeida, F. Role of Exosomal miRNAs and the Tumor Microenvironment in Drug Resistance. *Cells* **2020**, *9*, 1450. [CrossRef] [PubMed]

35. Chen, X.; Song, E. Turning foes to friends: Targeting cancer-associated fibroblasts. *Nat. Rev. Drug Discov.* **2019**, *18*, 99–115. [CrossRef] [PubMed]

36. Ma, L.; Young, J.; Prabhala, H.; Pan, E.; Mestdagh, P.; Muth, D.; Teruya-Feldstein, J.; Reinhardt, F.; Onder, T.T.; Valastyan, S.; et al. miR-9, a MYC/MYCN-activated microRNA, regulates E-cadherin and cancer metastasis. *Nat. Cell Biol.* **2010**, *12*, 247–256. [CrossRef]

37. D'Ippolito, E.; Plantamura, I.; Bongiovanni, L.; Casalini, P.; Baroni, S.; Piovan, C.; Orlandi, R.; Gualeni, A.V.; Gloghini, A.; Rossini, A.; et al. miR-9 and miR-200 Regulate PDGFRβ-Mediated Endothelial Differentiation of Tumor Cells in Triple-Negative Breast Cancer. *Cancer Res.* **2016**, *76*, 5562–5572. [CrossRef]

38. Zhuang, G.; Wu, X.; Jiang, Z.; Kasman, I.; Yao, J.; Guan, Y.; Oeh, J.; Modrusan, Z.; Bais, C.; Sampath, D.; et al. Tumour-secreted miR-9 promotes endothelial cell migration and angiogenesis by activating the JAK-STAT pathway. *EMBO J.* **2012**, *31*, 3513–3523. [CrossRef]

39. Cataldo, A.; Romero-Cordoba, S.; Plantamura, I.; Cosentino, G.; Hidalgo-Miranda, A.; Tagliabue, E.; Iorio, M.V. MiR-302b as a Combinatorial Therapeutic Approach to Improve Cisplatin Chemotherapy Efficacy in Human Triple-Negative Breast Cancer. *Cancers* **2020**, *12*, 2261. [CrossRef]

40. Hwang, C.-F.; Chien, C.-Y.; Huang, S.-C.; Yin, Y.-F.; Huang, C.-C.; Fang, F.-M.; Tsai, H.-T.; Su, L.-J.; Chen, C.-H. Fibulin-3 is associated with tumour progression and a poor prognosis in nasopharyngeal carcinomas and inhibits cell migration and invasion via suppressed AKT activity. *J. Pathol.* **2010**, *222*, 367–379. [CrossRef]

41. Almeida, M.; Costa, V.L.; Costa, N.R.; Ramalho-Carvalho, J.; Baptista, T.; Ribeiro, F.R.; Paulo, P.; Teixeira, M.R.; Oliveira, J.; Lothe, R.A.; et al. Epigenetic regulation of EFEMP1 in prostate cancer: Biological relevance and clinical potential. *J. Cell. Mol. Med.* **2014**, *18*, 2287–2297. [CrossRef] [PubMed]

42. Luo, R.; Zhang, M.; Liu, L.; Lu, S.; Zhang, C.Z.; Yun, J. Decrease of fibulin-3 in hepatocellular carcinoma indicates poor prognosis. *PLoS ONE* **2013**, *8*, e70511. [CrossRef] [PubMed]

43. Hu, Y.; Pioli, P.D.; Siegel, E.; Zhang, Q.; Nelson, J.; Chaturbedi, A.; Mathews, M.S.; Ro, D.I.; Alkafeef, S.; Hsu, N.; et al. EFEMP1 suppresses malignant glioma growth and exerts its action within the tumor extracellular compartment. *Mol. Cancer* **2011**, *10*, 123. [CrossRef] [PubMed]

44. Song, E.-L.; Hou, Y.-P.; Yu, S.-P.; Chen, S.-G.; Huang, J.-T.; Luo, T.; Kong, L.-P.; Xu, J.; Wang, H.-Q. EFEMP1 expression promotes angiogenesis and accelerates the growth of cervical cancer in vivo. *Gynecol. Oncol.* **2011**, *121*, 174–180. [CrossRef] [PubMed]

45. Camaj, P.; Seeliger, H.; Ischenko, I.; Krebs, S.; Blum, H.; De Toni, E.N.; Faktorova, D.; Jauch, K.-W.; Bruns, C.J. EFEMP1 binds the EGF receptor and activates MAPK and Akt pathways in pancreatic carcinoma cells. *Biol. Chem.* **2009**, *390*, 1293–1302. [CrossRef]

46. Yin, X.; Fang, S.; Wang, M.; Wang, Q.; Fang, R.; Chen, J. EFEMP1 promotes ovarian cancer cell growth, invasion and metastasis via activated the AKT pathway. *Oncotarget* **2016**, *7*, 47938–47953. [CrossRef]

47. Sadr-Nabavi, A.; Ramser, J.; Volkmann, J.; Naehrig, J.; Wiesmann, F.; Betz, B.; Hellebrand, H.; Engert, S.; Seitz, S.; Kreutzfeld, R.; et al. Decreased expression of angiogenesis antagonist EFEMP1 in sporadic breast cancer is caused by aberrant promoter methylation and points to an impact of EFEMP1 as molecular biomarker. *Int. J. Cancer* **2009**, *124*, 1727–1735. [CrossRef]

48. Noonan, M.M.; Dragan, M.; Mehta, M.M.; Hess, D.A.; Brackstone, M.; Tuck, A.B.; Viswakarma, N.; Rana, A.; Babwah, A.V.; Wondisford, F.E.; et al. The matrix protein Fibulin-3 promotes KISS1R induced triple negative breast cancer cell invasion. *Oncotarget* **2018**, *9*, 30034–30052. [CrossRef]

Review

Cancer Stem Cells: Devil or Savior—Looking behind the Scenes of Immunotherapy Failure

Lorenzo Castagnoli [1,†], Francesca De Santis [2,†], Tatiana Volpari [2], Claudio Vernieri [3,4], Elda Tagliabue [1], Massimo Di Nicola [2,†] and Serenella M. Pupa [1,*,†]

[1]   Department of Research, Molecular Targeting Unit, Fondazione IRCCS Istituto Nazionale dei Tumori di Milano, Via Amadeo 42, 20133 Milan, Italy; lorenzo.castagnoli@istitutotumori.mi.it (L.C.); elda.tagliabue@istitutotumori.mi.it (E.T.)

[2]   Department of Medical Oncology and Hematology, Unit of Immunotherapy and Anticancer Innovative Therapeutics, Fondazione IRCCS Istituto Nazionale dei Tumori di Milano, Via Venezian 1, 20133 Milan, Italy; francesca.desantis@istitutotumori.mi.it (F.D.S.); tatiana.volpari@istitutotumori.mi.it (T.V.); massimo.dinicola@istitutotumori.mi.it (M.D.N.)

[3]   Department of Medical Oncology and Hematology, FIRC Institute of Molecular Oncology, Fondazione IRCCS Istituto Nazionale dei Tumori, Via Venezian 1, 20133 Milan, Italy; claudio.vernieri@istitutotumori.mi.it

[4]   IFOM, FIRC Institute of Molecular Oncology, via Adamello 16, 20139 Milan, Italy

*   Correspondence: serenella.pupa@istitutotumori.mi.it; Tel.: +39-02-2390-2573; Fax: +39-02-2390-2692

†   These authors contributed equally to this work.

Received: 17 January 2020; Accepted: 21 February 2020; Published: 27 February 2020

Abstract: Although the introduction of immunotherapy has tremendously improved the prognosis of patients with metastatic cancers of different histological origins, some tumors fail to respond or develop resistance. Broadening the clinical efficacy of currently available immunotherapy strategies requires an improved understanding of the biological mechanisms underlying cancer immune escape. Globally, tumor cells evade immune attack using two main strategies: avoiding recognition by immune cells and instigating an immunosuppressive tumor microenvironment. Emerging data suggest that the clinical efficacy of chemotherapy or molecularly targeted therapy is related to the ability of these therapies to target cancer stem cells (CSCs). However, little is known about the role of CSCs in mediating tumor resistance to immunotherapy. Due to their immunomodulating features and plasticity, CSCs can be especially proficient at evading immune surveillance, thus potentially representing the most prominent malignant cell component implicated in primary or acquired resistance to immunotherapy. The identification of immunomodulatory properties of CSCs that include mechanisms that regulate their interactions with immune cells, such as bidirectional release of particular cytokines/chemokines, fusion of CSCs with fusogenic stromal cells, and cell-to-cell communication exerted by extracellular vesicles, may significantly improve the efficacy of current immunotherapy strategies. The purpose of this review is to discuss the current scientific evidence linking CSC biological, immunological, and epigenetic features to tumor resistance to immunotherapy.

Keywords: cancer stem cells; immunotherapy; tumor microenvironment; immune checkpoint blockade

---

1. Introduction

Consistent with the concept of cancer immunoediting, many pieces of evidence have underlined the existence of bidirectional crosstalk between cancer cells and cells of innate or adaptive immunity. Specifically, cancer immunoediting, which can constrain or promote tumor development and progression depending on the balance between cancer and immune cells, is a multistep process consisting of different and interchangeable scenarios: 1) the clearance of cancer cells by immune cells, 2) an equilibrium between cancer and immune cells, and 3) the escape phase, with a prevalence of cancer cells over immune cells [1]. During tumor progression, cancer cells acquire specific biological characteristics that lead to immune tolerance, thus preventing or hampering tumor cell attack and killing by antitumor immune cells [2]. In particular, the overexpression of inhibitory immune checkpoints, which impair the anticancer immune response, and/or the release of immunosuppressive cytokines/chemokines are the most common mechanisms that cancer cells utilize to protect themselves from the attack of cytotoxic immune cells [3]. In addition to these mechanisms, genomic instability [4], antigen (ag) loss or downregulation of the ag-presenting machinery [5], the generation of cell hybrids in the tumor microenvironment (TME) [6], the release of extracellular vesicles (EVs) as powerful mediators of intercellular communication [7], and the hierarchical tumor organization arising from cancer stem cells (CSCs) could contribute to immune escape in human cancers [8].

CSCs represent a minor subset of malignant cells capable of unlimited self-renewal and differentiation that contribute to tumorigenesis and tumor aggressiveness, tumor heterogeneity, metastasis, and resistance to antitumor therapies [9,10]. Through asymmetric cell division, a process that underlines the unlimited self-renewal capabilities of CSCs, a single CSC can hierarchically reconstitute the whole cancer cell population, thus regenerating/reseeding the original tumor if implanted in a different organism or in a different site of the same organism; this programme has been defined as "clonal tumor initiation capacity" [9,11]. The ability to shift between different phenotypic cell states by adapting their transcriptome to changes in the surrounding microenvironment confers CSCs the potential to transdifferentiate and invade other tissues and organs, a process also referred to as epithelial-mesenchymal transition (EMT) [12]. Moreover, while cytotoxic agents target the bulk of highly proliferating tumor cells, slowly cycling CSCs can resist chemotherapy and/or radiotherapy, finally resulting in aggressive/advanced treatment-refractory disease [13,14].

Recent studies suggest that CSCs could be crucial players in cancer immune escape; indeed, because of their immunomodulating properties, they can evade immunosurveillance and remain in a quiescent state, thus preventing lethal attack by antitumor immune cells [15–17]. Conversely, specific intratumor immune cell populations of the tumor niche interact with CSCs, thus affecting their functional status [18,19]. This biological crosstalk between CSCs and host immunity could represent a new "evil axis" responsible for primary or acquired tumor resistance to immunotherapy, thus paving the way for new therapeutic approaches based on the combination of anti-CSC treatments with immune checkpoint inhibitors (ICIs). In addition, cell–cell fusion, a process that under pathological conditions generates hybrids of tumor cells with diverse types of microenvironmental fusogenic cells, including bone marrow-derived and mesenchymal stem/multipotent stromal cells, macrophages, and fibroblasts, contributes to the formation of aberrant cells with tumor stem cell-like properties associated with tumor initiation, progression, and metastasis [6,20,21]. In general, cell fusion is a genetically regulated process, but external factors, such as hypoxia, inflammation, and mediators of intercellular communication, may also be involved in cell hybrid generation. In particular, it has been reported that EVs, including exosomes, may transport biological cargoes that could to some extent favor cell fusion and the formation of cancer hybrid stem cells that, in turn, promote tumor proliferation, immune evasion, and invasiveness [6]. Notably, EVs have also been found to exert a bidirectional role—EVs from CSCs to the TME and EVs from the TME to CSCs—in different solid cancers, such as breast, colon, lung, and prostate cancers, possibly promoting the development of premetastatic niches [7].

To further complicate this interactive scenario, epigenetic perturbations could also significantly contribute to CSC-related immune escape mechanisms [22]. Indeed, epigenetic modifications of both CSCs and differentiated cancer cells may lead to changes in the gene expression of immune-related genes, which can impact antigen processing/presentation and immune evasion. For instance, demethylating agents may enable re-expression of immune-related genes with potential for therapeutic impact, especially in the setting of combination treatment with immunotherapy [23].

In this review, we aim to highlight specific immunological traits of CSCs, pointing out how an immunosuppressive phenotype can promote tumor aggressiveness, stemness enrichment, and immunotherapy resistance.

2. CSCs: Leading Actors in the Intratumor Immunosuppressive Scenario?

The "CSC hypothesis" relies on the existence of a minor population of cancer cells, namely, CSCs, with the peculiar capability to undergo asymmetric divisions, which give rise to another CSC (self-renewal) and, at the same time, to a more differentiated precursor cell that recapitulates a tumor in all its components, including cells with different proliferative and invasive properties. Several studies have demonstrated that CSCs drive cancer cell proliferation, aggressiveness, and recurrence through the activation of specific signaling pathways, such as the Notch, Hedgehog, Wingless (WNT) and NF-kB circuits [22,24]. Notably, CSCs have also been implicated in primary or acquired tumor resistance to radiotherapy, chemotherapy, targeted therapies, and immunotherapy; in particular, CSCs contribute to tumor heterogeneity and metastasis [25]. For these reasons, CSCs may represent the "root of cancer", and their elimination represents a promising yet investigational therapeutic strategy. To date, the main limitation to the development of effective anti-CSC treatments has been the difficulty of identifying biological markers that univocally characterize CSCs when compared to their normal counterparts.

The interplay between CSCs and immune cells is an emerging field of investigation, and different studies have highlighted a key role of CSCs in "re-educating" the immune system, thus modifying the balance between antitumorigenic and protumorigenic immune cells [15–17]. The processing and presentation of tumor-associated ags (TAAs) in the tumor cell membrane are two crucial mechanisms driving tumor cell attack by immune cells [26]. TAAs are processed by the cytoplasmic proteasome machinery, transported into the endoplasmic reticulum by TAP1 and TAP2 proteins, loaded onto major histocompatibility complex I (MHC I), and presented to naive CD4+ helper and CD8+ cytotoxic T lymphocytes [27] (Figure 1). Different TAAs expressed by CSCs, such as ALDH1A1, CD133, CEP55, COA-1, EpCam, HEATR1 IL-13Rα2, SOX2, and DNAJB8, are recognized by T cells and are potentially capable of eliciting an effective antitumor immune response [28]. To prevent attack by immune cells, neoplastic cells in clinically detectable tumors often downregulate the expression of proteins involved in ag presentation, thus preventing the exposure of TAAs in tumor cell membranes [29]. This phenomenon appears to be enhanced in CSCs of several cancer types. In particular, a significant decrease in TAP2 and MHC I expression was revealed in the CSC compartment of head and neck and colorectal carcinomas, melanomas, and glioblastomas [30–33]. The molecular mechanisms responsible for the negative modulation of ag presentation in CSCs are still under investigation [31,33]. Even if the downregulation of members of the ag presentation machinery appears to be a common mechanism used by CSCs to avoid recognition by immune cells, it is not the only mechanism. For instance, studies performed in models of melanoma and glioma revealed no defective expression of HLA class I molecules [34,35]. In fact, CSCs can contribute to an immunosuppressive environment by upregulating the expression of immune checkpoint molecules that impair the activity of immune cells, such as CD200 and PD-L1 (Figure 1). CD200 is a protein expressed on the tumor cell surface, and it is able to promote immunosuppression [36,37]. Indeed, the binding of CD200 on tumor cells with its receptor (CD200R) expressed on T lymphocytes can induce a switch from an antitumor T helper (Th)-1 to an immunosuppressive Th-2 response [38]. CD200 overexpression has been reported in the CSCs of human papillomavirus (HPV)-positive and HPV-negative head and neck squamous cell carcinomas, thus pointing out its candidacy as a potentially actionable CSC marker [39] and paving the way for new therapeutic approaches targeting this immunomodulation-associated protein [40]. PD-L1, an immune checkpoint protein known to trigger a protumorigenic immunophenotype, was also found to be upregulated on the plasma membrane of melanoma as well as ovarian, breast, colon, and lung CSCs [41–44]. In particular, PD-L1 binding to its receptor (PD-1) on the T cell plasma membrane inhibits the production of immunostimulating cytokines (i.e., IFN-γ and IL-2) [45,46], dampens the T cell-mediated immune response [47], and promotes the expansion of immunosuppressive regulatory T (T-reg) cells [48].

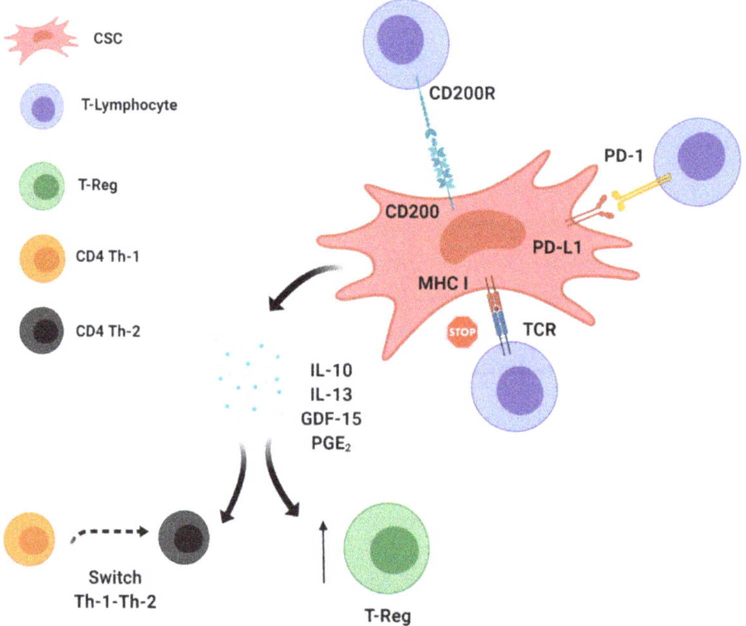

Figure 1. Schematic representing the core findings described in Section Altered expression of molecules involved in ag presentation (MHC I), the activity of immune cells (CD200 and PD-L1), and the release of immunomodulatory molecules with protumorigenic effects (IL-10, IL-13, GDF-15, and PGE$_2$) by CSCs. Created by Biorender.

The recent introduction of monoclonal antibodies (mAbs) targeting the PD-1/PD-L1 signaling axis in cancer therapy significantly improved the survival of patients with different advanced and aggressive tumor types, including melanoma, non-small-cell lung cancer, and kidney carcinoma [49]. In agreement with another group [50], we recently showed that the activation of the WNT signaling pathway increased PD-L1 expression on the plasma membrane of triple-negative breast CSCs (TNBCSCs) [42]. We also provided the first evidence that PD-L1+ TNBCSCs interact with tumor-infiltrating CD8+ and PD-1+ immune cells, thus halting antitumor immune responses [42].

In the complex process of immunoediting, a key role is played by cytokines released by the different cell subsets present in the TME. Depending on the type of molecule, different cytokines could promote the killing of cancer cells or, alternatively, they could contribute to establishing an immunosuppressive environment (Figure 1). CSCs can contribute to this mechanism by releasing cytokines that modulate the activity of tumor-infiltrating cells (Figure 1). For instance, melanoma and lung CSCs release IL-10 [32], a proinflammatory cytokine that exerts a negative regulation on T cells and stimulates the expansion of T-reg cells [51]. IL-13, a proinflammatory cytokine that impairs the cytotoxic activity of CD8+ T cells [52], was found to be overexpressed and released by lung CSCs [53]. Moreover, growth differentiation factor (GDF)-15, a member of the transforming growth factor-beta (TGF-β) superfamily of cytokines, was detected in colorectal carcinoma and glioma tumor specimens, where it contributed to promoting EMT, tumor metastasis, proliferation, and immune escape [54,55]. Additionally, GDF15 was found to maintain CSCs in breast cancer tissues by inducing its own expression in an autocrine/paracrine manner [56]. Furthermore, CSCs from glioma and colorectal neoplasms can also produce small molecules with immune-suppressive action. Specifically, the stemness-related WNT signaling pathway is involved in the production of prostaglandin (PG) E$_2$ [57], a derivative of arachidonic acid with the capability of inducing a shift from a Th-1 to a

Th-2 immune response, and its depletion has been shown to promote antitumor immune attack [58]. Together, these studies indicate the possibility of combining therapies directed against CSCs (by identifying and targeting CSC-related biomarkers) with immune-modulating agents that are already known to promote an effective antitumor immune response. Several ongoing phase I trials are testing the efficacy of WNT (NCT02521844, NCT02675946, and NCT02013154) or Hedgehog (NCT02690948) inhibitors in combination with pembrolizumab, a humanized MAb directed at PD-1.

3. Immunomodulatory Molecules Stimulating Tumor Stemness: The Dark Side of CSCs

CSCs are involved in bidirectional crosstalk with protumor or antitumor immune cells. Specific factors that are produced by CSCs to inhibit antitumor immunity also sustain CSC maintenance/activity, thus increasing tumor aggressiveness: a "direct mechanism" from CSCs to immune cells. Conversely, populations of tumor-infiltrating immune cells regulate CSCs by establishing physical interactions with them or through the release of soluble factors, such as cytokines: a "reverse mechanism" from immune cells to CSCs (Figure 2).

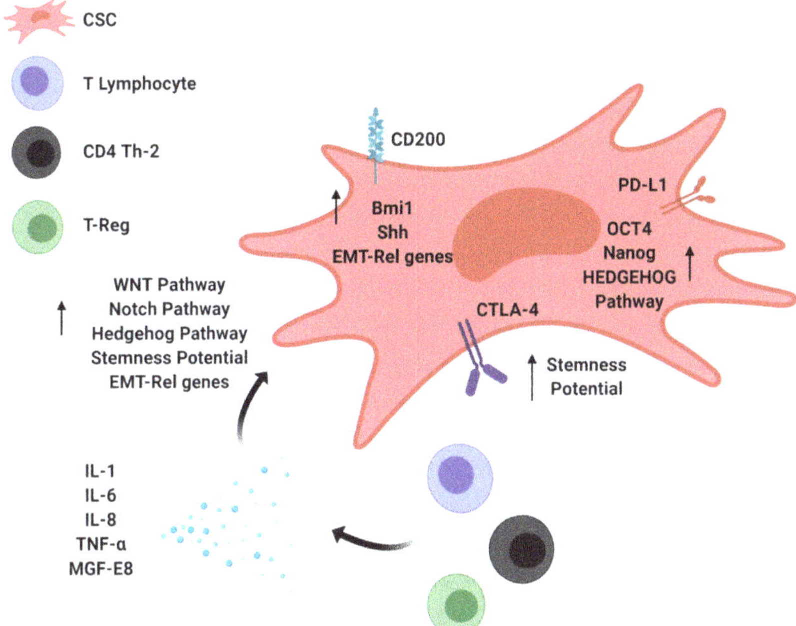

Figure 2. Schematic representing the core findings described in Section Altered expression of molecules involved in intratumor stemness enrichment (CD200, PD-L1, and CTLA-4) occurs via a "direct mechanism". Dynamic crosstalk occurs between cancer stem cells (CSCs) and immune cells through the release of inflammatory soluble factors (IL-1, IL-6, IL-8, TNF-α, and MGF-E8) by tumor-infiltrating immune cells—a "reverse mechanism"—coupled with the activation of CSC-related signaling pathways (WNT, Notch, and Hedgehog) that sustain the epithelial-mesenchymal transition (EMT) program and CSC maintenance/activity. Created byBiorender.

Regarding the "direct mechanism", CD200 expression by CSCs not only inhibits immune system activation (see the previous section) but also sustains stemness enrichment by stimulating the expression of Bmi1 and Shh, two molecules implicated in Hedgehog signaling [59,60] (Figure 2). Moreover, ectopic CD200 expression induces EMT-related genes, enhanced resistance to radiotherapy, and invasiveness in head and neck squamous cell carcinoma and skin tumors [61,62].

PD-L1 expression by cancer cells has also been associated with enhanced stemness potential in different oncotypes. In particular, Almozyan S et al. demonstrated that PD-L1 expression in breast cancer was associated with resistance to chemotherapy and with the induction of EMT [63]. In this study, the authors revealed that PD-L1-silenced breast cancer cells formed mammospheres with significantly decreased efficiency and displayed impaired in vivo tumor-forming capacity when injected in mice in limiting dilution conditions, thus providing evidence for a regulatory role of PD-L1 in CSC self-renewal [63]. In addition, the same authors also reported a direct role of PD-L1 in controlling the expression of the stemness-related factors OCT4, Nanog, and BMI1 [63] (Figure 2). Together, these data indicate that PD-L1 expression in cancer cells not only inhibits antitumor immunity but also may contribute to the maintenance of stemness potential. Consistent with this model, Wei F et al. demonstrated the existence of crosstalk between PD-L1 and the Hedgehog signaling pathway that is mediated through a direct interaction between PD-L1 and HMGA1 [64]. In this study, PD-L1 increased the frequency of intratumor CSCs and contributed to the maintenance of tumor cell self-renewal [64]. Additionally, CTLA4, another immune checkpoint protein that inhibits T cell proliferation, activation, and anticancer immune response [65], was shown to increase the stemness potential in melanoma cells [66] (Figure 2). Indeed, CTLA-4 has a specific role in maintaining stem cell activity, and its inhibition with specific anti-CTLA-4 mAbs suppressed crucial CSC properties, including sphere-forming capability and in vivo tumorigenic potential [66].

Regarding the "reverse mechanism", soluble inflammatory cytokines released into the TME by tumor-infiltrating immune cells could modulate CSC activities as well. For instance, the soluble factor IL-1 promoted Th-1 to Th-2 differentiation, thus contributing to the inhibition of antitumor immunity [67,68]. In particular, IL-1β stimulated the expression of the CSC-related marker ZEB1, induced resistance to chemotherapy, and increased sphere-forming efficiency in colon cancer cells, thus supporting a role for IL-1β in promoting CSC self-renewal and EMT [69]. In pancreatic cancer, KRAS-driven signaling activates an intrinsic inflammatory programme by activating the CSC-related NF-κB signaling pathway, which in turn stimulates the production of cytokines of the IL-1 family that promote early tumor invasiveness, EMT, angiogenesis, and metastasis [70]. Other proinflammatory cytokines, such as IL-6 and IL-8, could increase the number and activation of specific CSC subpopulations. In particular, IL-6 can act intrinsically on tumor cells, but it can also stimulate antitumor adaptive immunity [71]. IL-6 stimulates the expression of genes involved in stemness, invasion, migration, and tumorigenesis through a mechanism that involves its interaction with STAT3 [72,73]. IL-6 also causes resistance to chemotherapy, a process strongly linked to CSC enrichment, in different tumor types, including renal and ovarian carcinomas [74,75]. IL-8, a proinflammatory cytokine able to convert the anticancer immune response into an immunosuppressive response through its binding with CXCR1/2 receptors [76], is also involved in the acquisition and/or maintenance of stemness features in lung, renal, pancreatic, and breast cancers [77,78]. In particular, the inhibition of the IL-8/CXCR1/2 axis efficiently targets the CSC compartment in pancreatic cancer [78], and the combination of repertaxin, an antagonist of the CXCR1 receptor, with lapatinib, a tyrosine kinase inhibitor that halts HER2-mediated signaling, impairs mammosphere formation by HER2-positive cancer cells more efficiently than repertaxin or lapatinib monotherapy [79]. In this complex scenario, tumor necrosis factor (TNF)-α, a potent proinflammatory cytokine with putative suppressive effects in antitumor immunity and that has been reported to be involved in the expansion of T-reg cells [80], is implicated in the regulation of CSC-related features. In particular, TNF-α promotes cancer invasion and metastasis associated with the EMT programme in colorectal cancer [81]. Furthermore, milk fat globule epidermal growth factor-8 (MGF-E8), a secreted mediator of local immune suppression that stimulates T-reg cell infiltration and suppresses the Th-1 response and NK cell and CD8+ T cell cytotoxicity [82], was also found to enhance tumor cell survival, invasion, and angiogenesis and to modulate CSC activity. Specifically, in colon cancer, MGF-E8 released by TAMs induces the activation of the Hedgehog pathway, driving melanoma progression [83]. All these considerations indicate that an immunosuppressive TME not only affects the function of intratumor immune cells but also has an impact on CSC number and activation status.

4. Epigenetic Control of Stemness Features and Immune Escape

Among the mechanisms regulating stemness properties, epigenetic mechanisms play a pivotal role by shaping the plasticity and biological behaviour of CSCs. Epigenetics refers to the set of dynamic changes in gene expression and cellular phenotypes that occur in the absence of genomic modifications. In recent years, the study of epigenetics has attracted much interest because epigenetic modifications have been shown to heavily affect tumor cell fate [84]. The main epigenetic events that contribute to cancer development consist of aberrant DNA methylation, histone modifications, chromatin remodeling, and changes in noncoding RNAs, including miRNAs. In particular, the discovery of histone post-translational (hPT) modifications led to the so-called "histone code hypothesis", according to which a dynamic constellation of biochemical modifications determines the binding of chromatin-remodeling factors to the nucleosome, a region of DNA wrapped around eight histone protein cores [85,86]. By altering chromatin structure, chromatin-remodeling factors regulate DNA accessibility by transcription factors, co-factors, and the general transcription machinery, thus modulating gene expression in cancer cells. These biological processes govern the epigenome landscape, which represents a central element during cancer progression and the expression of stemness properties. In this context, breakthrough studies by McCullough and Weismann have described the epigenetic-driven hierarchical organization of the hematopoietic system, with leukemia representing a paradigm governed by plastic leukemic stem cells [87,88]. Specifically, different studies addressing the link between stemness phenotypes and epigenetic control have analyzed the formation of the LSC compartment in acute lymphoblastic leukemia and acute myeloid leukemia. These studies highlighted the presence of chromosomal rearrangements involving KTM2A/MLL, which encodes a modified histone methyltransferase that is able to constitutively activate DNA transcription. It is evident how the establishment of epigenetic aberrations serves as an early event leading to the accumulation of cells with increased stemness properties and self-renewal ability.

Among brain neoplasms, glioblastoma represents the most aggressive and deadly type and is characterized by high histologic grade, an undifferentiated phenotype, and a high frequency of CSCs. Several studies have identified gain-of-function mutations in genes encoding histone H3 in pediatric high-grade gliomas, with at least two recurrent mutations within the histone H3.3 gene H3F3A identified in nearly 50% of the patients [89,90]. These two mutations result in the substitutions of the K27 or G34 residue with a methionine (K27M), an arginine (G34R), or a valine (G34V) residue. Of note, the K27M substitution causes the inhibition of polycomb repressive complex 2, with a genome-wide reduction in H3K27me3 and the re-establishment of an earlier developmental program in neural precursor cells and the reacquisition of oncogenic self-renewal ability [91].

In breast cancer, key CSC-related gene pathways are finely regulated by epigenetic mechanisms and their inhibition affects cancer cell survival, self-renewal, tumorigenesis, and progression [92]. For instance, epigenetic modifications regulate genes of the Notch and WNT pathways, which control normal mammary development and breast cancer self-renewal [93]. Moreover, nuclear receptor corepressor 2, which recruits histone deacetylases (HDACs) to the promoter regions of Notch target genes (e.g., MYC and HES1) to regulate their transcription [94], has been found to be upregulated as a result of reduced recruitment of corepressor complexes with chromatin-remodeling activity, finally resulting in enhanced transcription of Notch target genes [95]. In a variety of human solid malignancies, including colorectal cancer, hepatocellular cancer, and breast cancer [96], evidence is accumulating that epigenetic mechanisms promote tumorigenesis by leading to the activation of the WNT/β-catenin pathway. For instance, DNA methylation and histone modifications of WNT regulators, such as WNT inhibitor factor 1, AXIN2, secreted frizzled-related protein 1, and Dickkopf-related protein 1, have been frequently reported in breast cancer and are responsible for aberrant WNT/β-catenin gene pathway expression/activation [97]. Moreover, some WNT proteins, such as WNT1, WNT2, WNT3A, and WNT5A, which are frequently overexpressed in breast, colon, lung, and prostate cancers as a result of epigenetic modifications, behave as oncogenic activators of the canonical WNT signaling pathway [98]. Because aberrant activation of the WNT/β-catenin axis contributes to cancer progression, the identification of epigenetic events associated with WNT/β-catenin activation could provide useful biomarkers for cancer detection and prognosis.

Specific epigenetic events taking part in tumor progression are also involved in tumor immune escape. For instance, different studies, reviewed in [99], have identified mutations or structural defects in human and mouse genes that are known to be epigenetically regulated in CSCs and are involved in the control of T and B cell differentiation [100–102]. Moreover, epigenetic modulators have been shown to restore functional immune cell recognition and immunogenicity, providing a proof-of-concept demonstration that targeting tumor epigenetics could have synergistic antitumor activity with currently available immunotherapy strategies (Figure 3) [103–105].

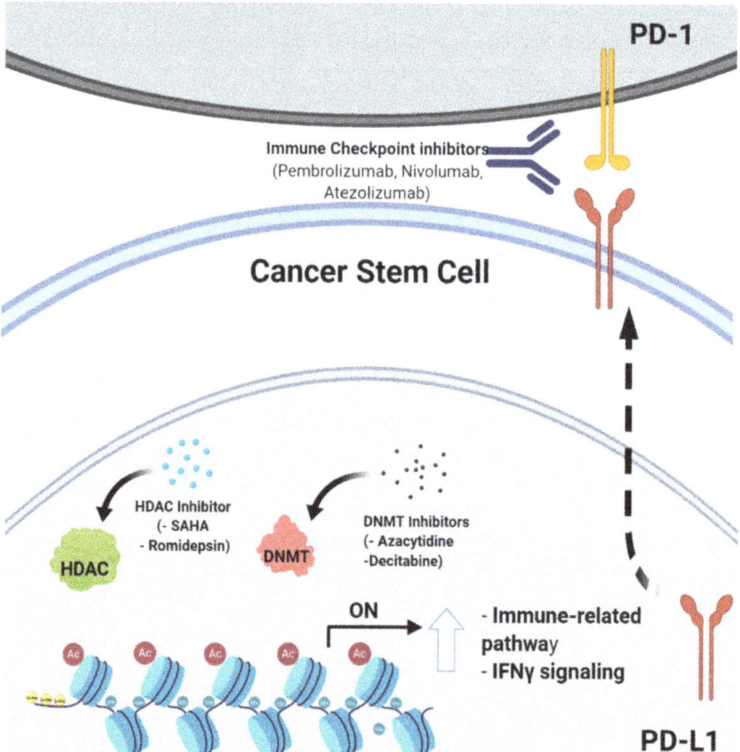

Figure 3. Schematic representing several epigenetic modulators (histone deacetylases (HDAC) and DNA methyltransferase (DNMT) inhibitors) that may act by sensitizing tumor cells to immune checkpoint blockade. Created by Biorender.

Taken together, published data indicate that bulk tumor cells and CSCs may exploit epigenetic repression of specific immune genes to escape killing by immune cells. It is likely that the interplay between genetic and epigenetic modifications affects the carcinogenesis process and takes part in the selective pressures involved in immune cell escape. This has been clearly demonstrated in an analysis of several melanoma cell lines derived from patients who underwent successful immunotherapy and recurrence [106]. The scientific rationale for the use of HDAC inhibitors (HDACi) or DNA methyltransferase inhibitors (DNMTi) in ongoing clinical trials in cancer patients relies on the ability of these drugs to inhibit tumor cell growth and to induce cell differentiation [107]. However, DNMTi and HDACi are also capable of upregulating the expression of TAAs in different solid tumors [108–111], thus potentially sensitizing cancer cells to ICIs by upregulating CTLA4, PD-1, PD-L1, and PD-L2 molecules on both tumor cells and TILs [112,113]. HDACi can also restore the expression of MHC class I and II molecules, which are downregulated as a result of transcriptional alterations in the IFNγ and NF-kB (for MHC class I) or the CIITA (for MHC class II) pathways [108], thus facilitating innate and adaptive antitumor immune responses. Other studies have linked HDACi to both increased and decreased expression of PD-L1 and PD-L2 on tumor cells [114,115]. However, the great potential for epigenetic-targeted interventions lies in the fact that, unlike genetic abnormalities, epigenetic modifications are reversible, thus allowing fast and potentially complete functional restoration of otherwise intact genes [113,116]. An exciting advance towards the combination of epigenetic and immune therapies is the fact that epigenetic therapy may reverse immune tolerance in human cancers [115,116]. Clinical trials are ongoing to explore this hypothesis. This kind of combination therapy may significantly benefit from the advent of next-generation sequencing (NGS)-based approaches, which massively amplify the possibility of studying new strategies, with the goal of achieving successful precision medicine and overcoming resistance to immunotherapy [117]. In particular, these techniques are based on the analysis of the epigenome, defined by a set of chemical modifications, such as methylation and acetylation of DNA and/or DNA-binding histone proteins, and include chromatin immunoprecipitation (ChIP) assays coupled to NGS, commonly known as ChIP-seq, and methylation analysis through whole-genome bisulfite/array-based sequencing. By harnessing the power of NGS, it is possible to analyze genome-wide DNA-binding sites for transcription factors and histone proteins and methylome patterns at a single-nucleotide resolution [118].

5. Conclusions

In this review, we have discussed the growing body of current scientific evidence linking CSC biological, immunological, and epigenetic features to tumor resistance to immunotherapy. A better understanding of the molecular mechanisms regulating the dynamic interplay between cells in the TME and CSCs could pave the way for future anticancer therapies aimed at targeting CSCs by reversing intratumor immunosuppression.

Author Contributions: Conceptualization, L.C., F.D.S., M.D.N., and S.M.P.; writing—original draft preparation, L.C., F.D.S., M.D.N., and S.M.P.; reviewing and editing, C.V., E.T., and S.M.P.; L.C. drew Figures 1 and 2; T.V. drew Figure. All authors have read and agreed to the published version of the manuscript.

Funding: This work was funded by grants from Ricerca Strategica Istituzionale call 2016 to M. Di Nicola and the Associazione Italiana Ricerca Cancro (AIRC) call 2015 Id. 16918 to S.M. Pupa.

Conflicts of Interest: The authors declare no conflicts of interest.

References

1. O'Donnell, J.S.; Teng, M.W.L.; Smyth, M.J. Cancer immunoediting and resistance to T cell-based immunotherapy. *Nat. Rev. Clin. Oncol.* **2018**, *16*, 151–167. [CrossRef] [PubMed]
2. Gonzalez, H.; Hagerling, C.; Werb, Z. Roles of the immune system in cancer: From tumor initiation to metastatic progression. *Genome Res.* **2018**, *32*, 1267–1284. [CrossRef] [PubMed]
3. Gun, S.Y.; Lee, S.W.L.; Sieow, J.L.; Wong, S.C. Targeting immune cells for cancer therapy. *Redox Boil.* **2019**, *25*, 101174. [CrossRef] [PubMed]
4. McGranahan, N.; Favero, F.; De Bruin, E.C.; Birkbak, N.J.; Szallasi, Z.; Swanton, C. Clonal status of actionable driver events and the timing of mutational processes in cancer evolution. *Sci. Transl. Med.* **2015**, *7*, 283ra54. [CrossRef] [PubMed]
5. Quail, D.F.; Joyce, J.A. Microenvironmental regulation of tumor progression and metastasis. *Nat. Med.* **2013**, *19*, 1423–1437. [CrossRef] [PubMed]
6. Jiang, E.; Yan, T.; Xu, Z.; Shang, Z. Tumor Microenvironment and Cell Fusion. *BioMed Res. Int.* **2019**, *2019*, 5013592-12. [CrossRef]

7.   Ciardiello, C.; Leone, A.; Budillon, A. The Crosstalk between Cancer Stem Cells and Microenvironment Is Critical for Solid Tumor Progression: The Significant Contribution of Extracellular Vesicles. *Stem Cells Int.* **2018**, *2018*, 1–11. [CrossRef]

8.   Greaves, M.; Mailey, C.C. Clonal evolution in cancer. *Nature* **2012**, *481*, 306–313. [CrossRef]

9.   Kreso, A.; Dick, J.E. Evolution of the Cancer Stem Cell Model. *Cell Stem Cell* **2014**, *14*, 275–291. [CrossRef]

10.   Batlle, E.; Clevers, H. Cancer stem cells revisited. *Nat. Med.* **2017**, *23*, 1124–1134. [CrossRef]

11.   Shackleton, M.; Vaillant, F.; Simpson, K.J.; Stingl, J.; Smyth, G.K.; Asselin-Labat, M.-L.; Wu, L.; Lindeman, G.J.; Visvader, J.E. Generation of a functional mammary gland from a single stem cell. *Nature* **2006**, *439*, 84–88. [CrossRef] [PubMed]

12.   Suzuki, M.; Mose, E.S.; Montel, V.; Tarin, D. Dormant Cancer Cells Retrieved from Metastasis-Free Organs Regain Tumorigenic and Metastatic Potency. *Am. J. Pathol.* **2006**, *169*, 673–681. [CrossRef] [PubMed]

13.   Phi, L.T.H.; Sari, I.N.; Yang, Y.-G.; Lee, S.-H.; Jun, N.; Kim, K.S.; Lee, Y.K.; Kwon, H.Y. Cancer Stem Cells (CSCs) in Drug Resistance and their Therapeutic Implications in Cancer Treatment. *Stem Cells Int.* **2018**, *2018*, 1–16. [CrossRef] [PubMed]

14.   Ayob, A.Z.; Ramasamy, T.S. Cancer stem cells as key drivers of tumour progression. *J. Biomed. Sci.* **2018**, *25*, 20. [CrossRef]

15.   Maccalli, C.; Rasul, K.I.; Elawad, M.; Ferrone, S. The role of cancer stem cells in the modulation of anti-tumor immune responses. *Semin. Cancer Boil.* **2018**, *53*, 189–200. [CrossRef]

16.   Boyd, A.S.; Rodrigues, N.P. Stem Cells Cycle toward Immune Surveillance. *Immuniy* **2018**, *48*, 187–190. [CrossRef]

17.   Miranda, A.; Hamilton, P.T.; Zhang, A.W.; Pattnaik, S.; Becht, E.; Mezheyeuski, A.; Bruun, J.; Micke, P.; De Reynies, A.; Nelson, B.H. Cancer stemness, intratumoral heterogeneity, and immune response across cancers. *Proc. Natl. Acad. Sci. USA* **2019**, *116*, 9020–9029. [CrossRef]

18.   Korkaya, H.; Liu, S.; Wicha, M.S. Regulation of cancer stem cells by cytokine networks: Attacking cancer's inflammatory roots. *Clin. Cancer Res.* **2011**, *17*, 6125–6129. [CrossRef]

19.   Sultan, M.; Coyle, K.; Vidovic, D.; Thomas, M.L.; Gujar, S.; Marcato, P. Hide-and-seek: The interplay between cancer stem cells and the immune system. *Carcinogenesis* **2016**, *38*, 107–118. [CrossRef]

20.   Lu, X.; Kang, Y. Cell Fusion Hypothesis of the Cancer Stem Cell. *Adv. Exp. Med. Biol.* **2011**, *950*, 129–140.

21.   Nagler, C.; Zänker, K.S.; Dittmar, T. Cell Fusion, Drug Resistance and Recurrence CSCs. *Adv. Exp. Med. Biol.* **2011**, *950*, 173–182.

22.   Pires, B.; E Souza, L.D.; Rodrigues, J.A.; De Amorim, Í.S.S.; Mencalha, A.L. Targeting Cellular Signaling Pathways in Breast Cancer Stem Cells and its Implication for Cancer Treatment. *Anticancer. Res.* **2016**, *36*, 5681–5692. [CrossRef] [PubMed]

23.   Seliger, B.; Hu-Lieskovan, S.; Wargo, J.A.; Ribas, A. Primary, Adaptive, and Acquired Resistance to Cancer Immunotherapy. *Cell* **2017**, *168*, 707–723.

24.   Ebben, J.D.; Treisman, D.M.; Zorniak, M.; Kutty, R.G.; Clark, P.A.; Kuo, J.S. The cancer stem cell paradigm: A new understanding of tumor development and treatment. *Expert Opin. Ther. Targets* **2010**, *14*, 621–632. [CrossRef] [PubMed]

25.   Shibue, T.; Weinberg, R.A. EMT, CSCs, and drug resistance: The mechanistic link and clinical implications. *Nat. Rev. Clin. Oncol.* **2017**, *14*, 611–629. [CrossRef] [PubMed]

26.   Sanchez-Paulete, A.R.; Teijeira, A.; Cueto, F.J.; Garasa, S.; Perez-Gracia, J.L.; Sanchez-Arraez, A.; Sancho, D.; Melero, I. Antigen cross-presentation and T-cell cross-priming in cancer immunology and immunotherapy. *Ann. Oncol.* **2017**, *28*, xii74. [CrossRef]

27.   Cresswell, P.; Bangia, N.; Dick, T.; Diedrich, G. The nature of the MHC class I peptide loading complex. *Immunol. Rev.* **1999**, *172*, 21–28. [CrossRef]

28.   Rasool, S.; Rutella, S.; Ferrone, S.; Maccalli, C. Cancer Stem Cells: The Players of Immune Evasion from Immunotherapy. In *Cancer Stem Cell Resistance to Targeted Therapy*; Maccalli, C., Todaro, M., Ferrone, S., Eds.; Springer Nature: Basel, Switzerland, 2019; pp. 223–249.

29.   Rodríguez, J.A. HLA-mediated tumor escape mechanisms that may impair immunotherapy clinical outcomes via T-cell activation. *Oncol. Lett.* **2017**, *14*, 4415–4427. [CrossRef]

30.   Chikamatsu, K.; Takahashi, G.; Sakakura, K.; Ferrone, S.; Masuyama, K. Immunoregulatory properties of CD44+ cancer stem-like cells in squamous cell carcinoma of the head and neck. *Head Neck* **2011**, *33*, 208–215. [CrossRef]

31. Volonté, A.; Di Tomaso, T.; Spinelli, M.; Todaro, M.; Sanvito, F.; Albarello, L.; Bissolati, M.; Ghirardelli, L.; Orsenigo, E.; Ferrone, S.; et al. Cancer-initiating cells from colorectal cancer patients escape from t cell–mediated immunosurveillance in vitro through membrane-bound IL-4. *J. Immunol.* **2014**, *192*, 523–532.

32. Schatton, T.; Schütte, U.; Frank, N.Y.; Zhan, Q.; Hoerning, A.; Robles, S.C.; Zhou, J.; Hodi, F.S.; Spagnoli, G.C.; Murphy, G.F.; et al. Modulation of T-cell activation by malignant melanoma initiating cells. *Cancer Res.* **2010**, *70*, 697–708. [CrossRef] [PubMed]

33. Di Tomaso, T.; Mazzoleni, S.; Wang, E.; Sovena, G.; Clavenna, D.; Franzin, A.; Mortini, P.; Ferrone, S.; Doglioni, C.; Marincola, F.M.; et al. Immunobiological characterization of cancer stem cells isolated from glioblastoma patients. *Clin. Cancer Res.* **2010**, *16*, 800–813. [CrossRef]

34. Pietra, G.; Manzini, C.; Vitale, M.; Balsamo, M.; Ognio, E.; Boitano, M.; Queirolo, P.; Moretta, L.; Mingari, M.C. Natural killer cells kill human melanoma cells with characteristics of cancer stem cells. *Int. Immunol.* **2009**, *21*, 793–801. [CrossRef] [PubMed]

35. Castriconi, R.; Daga, A.; Dondero, A.; Zona, G.; Poliani, P.L.; Melotti, A.; Griffero, F.; Marubbi, D.; Spaziante, R.; Bellora, F.; et al. NK Cells Recognize and Kill Human Glioblastoma Cells with Stem Cell-Like Properties. *J. Immunol.* **2009**, *182*, 3530–3539. [CrossRef]

36. Gorczynski, R.; Yu, K.; Clark, D. Receptor engagement on cells expressing a ligand for the tolerance-inducing molecule OX2 induces an immunoregulatory population that inhibits alloreactivity in vitro and in vivo. *J. Immunol.* **2000**, *165*, 4854–4860. [CrossRef] [PubMed]

37. Gorczynski, R.; Chen, Z.; Kai, Y.; Lee, L.; Wong, S.; Marsden, P.A. CD200 is a ligand for all members of the CD200R family of immunoregulatory molecules. *J. Immunol.* **2004**, *172*, 7744–7749. [CrossRef]

38. Kretz-Rommel, A.; Qin, F.; Dakappagari, N.; Ravey, E.P.; McWhirter, J.; Oltean, D.; Frederickson, S.; Maruyama, T.; Wild, M.A.; Nolan, M.-J.; et al. CD200 expression on tumor cells suppresses antitumor immunity: New approaches to cancer immunotherapy. *J. Immunol.* **2007**, *178*, 5595–5605. [CrossRef]

39. Jung, Y.-S.; Vermeer, P.D.; Vermeer, D.W.; Lee, S.-J.; Goh, A.R.; Ahn, H.-J.; Lee, J.H. CD200: Association with cancer stem cell features and response to chemoradiation in head and neck squamous cell carcinoma. *Head Neck* **2014**, *37*, 327–335. [CrossRef]

40. Deonarain, M.P.; Kousparou, C.A.; Epenetos, A.A. Antibodies targeting cancer stem cells: A new paradigm in immunotherapy? *MAbs* **2009**, *1*, 12–25. [CrossRef]

41. Gupta, H.B.; A Clark, C.; Yuan, B.; Sareddy, G.; Pandeswara, S.; Padron, A.S.; Hurez, V.; Conejo-Garcia, J.; Vadlamudi, R.; Li, R.; et al. Tumor cell-intrinsic PD-L1 promotes tumor-initiating cell generation and functions in melanoma and ovarian cancer. *Signal Transduct. Target. Ther.* **2016**, *1*, 16030. [CrossRef]

42. Castagnoli, L.; Cancila, V.; Cordoba-Romero, S.L.; Faraci, S.; Talarico, G.; Belmonte, B.; Iorio, M.; Milani, M.; Volpari, T.; Chiodoni, C.; et al. WNT signaling modulates PD-L1 expression in the stem cell compartment of triple-negative breast cancer. *Oncogene* **2019**, *38*, 4047–4060. [CrossRef] [PubMed]

43. Wu, Y.; Chen, C.; Xu, Z.P.; Gu, W. Increased PD-L1 Expression in Breast and Colon Cancer Stem Cells. *Clin. Exp. Pharmacol. Physiol.* **2017**, *44*, 602–604. [CrossRef] [PubMed]

44. Raniszewska, A.; Polubiec-Kownacka, M.; Rutkowska, E.; Domagala-Kulawik, J. PD-L1 Expression on Lung Cancer Stem Cells in Metastatic Lymph Nodes Aspirates. *Stem Cell Rev. Rep.* **2018**, *15*, 324–330. [CrossRef] [PubMed]

45. Amarnath, S.; Mangus, C.W.; Wang, J.C.M.; Wei, F.; He, A.; Kapoor, V.; Foley, J.E.; Massey, P.R.; Felizardo, T.C.; Riley, J.; et al. The PDL1-PD1 Axis Converts Human TH1 Cells into Regulatory T Cells. *Sci. Transl. Med.* **2011**, *3*, 111ra120. [CrossRef] [PubMed]

46. Francisco, L.M.; Sage, P.T.; Sharpe, A.H. The PD-1 pathway in tolerance and autoimmunity. *Immunol. Rev.* **2010**, *236*, 219–242. [CrossRef]

47. Jiang, X.; Wang, J.; Deng, X.; Xiong, F.; Ge, J.; Xiang, B.; Wu, X.; Ma, J.; Zhou, M.; Li, X.; et al. Role of the tumor microenvironment in PD-L1/PD-1-mediated tumor immune escape. *Mol. Cancer* **2019**, *18*, 10. [CrossRef]

48. Gianchecchi, E.; Fierabracci, A. Inhibitory Receptors and Pathways of Lymphocytes: The Role of PD-1 in Treg Development and Their Involvement in Autoimmunity Onset and Cancer Progression. *Front. Immunol.* **2018**, *9*, 2374. [CrossRef]

49. Alsaab, H.O.; Sau, S.; Alzhrani, R.; Tatiparti, K.; Bhise, K.; Kashaw, S.K.; Iyer, A.K. PD-1 and PD-L1 Checkpoint Signaling Inhibition for Cancer Immunotherapy: Mechanism, Combinations, and Clinical Outcome. *Front. Pharmacol.* **2017**, *8*, 8. [CrossRef]

50. Hsu, J.-M.; Xia, W.; Hsu, Y.-H.; Chan, L.-C.; Yu, W.-H.; Cha, J.-H.; Chen, C.-T.; Liao, H.-W.; Kuo, C.-W.; Khoo, K.-H.; et al. STT3-dependent PD-L1 accumulation on cancer stem cells promotes immune evasion. *Nat. Commun.* **2018**, *9*, 1908. [CrossRef]

51. Arce-Sillas, A.; Álvarez-Luquín, D.D.; Tamaya-Domínguez, B.; Gomez-Fuentes, S.; Trejo-García, A.; Melo-Salas, M.; Cárdenas, G.; Rodríguez-Ramírez, J.; Adalid-Peralta, L. Regulatory T Cells: Molecular Actions on Effector Cells in Immune Regulation. *J. Immunol. Res.* **2016**, *2016*, 1–12. [CrossRef]

52. Wynn, T.A. IL-13 effector functions. *Annu. Rev. Immunol.* **2003**, *21*, 425–456. [CrossRef] [PubMed]

53. Maccalli, C.; Volontè, A.; Cimminiello, C.; Parmiani, G. Immunology of cancer stem cells in solid tumours. A review. *Eur. J. Cancer* **2014**, *50*, 649–655. [CrossRef] [PubMed]

54. Li, C.; Wang, J.; Kong, J.; Tang, J.; Wu, Y.; Xu, E.; Zhang, H.; Lai, M. GDF15 promotes EMT and metastasis in colorectal cancer. *Oncotarget* **2015**, *7*, 860–872. [CrossRef] [PubMed]

55. Roth, P.; Junker, M.; Tritschler, I.; Mittelbronn, M.; Dombrowski, Y.; Breit, S.N.; Tabatabai, G.; Wick, W.; Weller, M.; Wischhusen, J. GDF-15 Contributes to Proliferation and Immune Escape of Malignant Gliomas. *Clin. Cancer Res.* **2010**, *16*, 3851–3859. [CrossRef]

56. Sasahara, A.; Tominaga, K.; Nishimura, T.; Yano, M.; Kiyokawa, E.; Noguchi, M.; Noguchi, M.; Kanauchi, H.; Ogawa, T.; Minato, H.; et al. An autocrine/paracrine circuit of growth differentiation factor (GDF) 15 has a role for maintenance of breast cancer stem-like cells. *Oncotarget* **2017**, *8*, 24869–24881. [CrossRef]

57. Goessling, W.; North, T.E.; Loewer, S.; Lord, A.M.; Lee, S.; Stoick-Cooper, C.L.; Weidinger, G.; Puder, M.; Daley, G.Q.; Moon, R.; et al. Genetic Interaction of PGE2 and Wnt Signaling Regulates Developmental Specification of Stem Cells and Regeneration. *Cell* **2009**, *136*, 1136–1147. [CrossRef]

58. Sakata, D.; Yao, C.; Narumiya, S. Emerging roles of prostanoids in T cell-mediated immunity. *IUBMB Life* **2010**, *62*, 591–596. [CrossRef]

59. Wang, X.; Venugopal, C.; Manoranjan, B.; McFarlane, N.; O'Farrell, E.; Nolte, S.; Gunnarsson, T.; Hollenberg, R.; Kwiecien, J.; Northcott, P.; et al. Sonic hedgehog regulates Bmi1 in human medulloblastoma brain tumor-initiating cells. *Oncogene* **2011**, *31*, 187–199. [CrossRef]

60. Carballo, G.B.; Honorato, J.R.; De Lopes, G.P.F.; Spohr, T.C.L.D.S.E. A highlight on Sonic hedgehog pathway. *Cell Commun. Signal.* **2018**, *16*, 11. [CrossRef]

61. Shin, S.P.; Goh, A.R.; Kang, H.G.; Kim, S.J.; Kim, J.K.; Kim, K.T.; Lee, J.H.; Bae, Y.S.; Jung, Y.S.; Lee, S.J. CD200 Induces Epithelial-to-Mesenchymal Transition in Head and Neck Squamous Cell Carcinoma via β-Catenin-Mediated Nuclear Translocation. *Cancers* **2019**, *11*, 1583. [CrossRef]

62. Stumpfova, M.; Ratner, D.; Desciak, E.B.; Eliezri, Y.D.; Owens, D.M. The immunosuppressive surface ligand CD200 augments the metastatic capacity of squamous cell carcinoma. *Cancer Res.* **2010**, *70*, 2962–2972. [CrossRef] [PubMed]

63. Almozyan, S.; Colak, D.; Mansour, F.; Alaiya, A.; Al-Harazi, O.; Qattan, A.; Al-Mohanna, F.; Al-Alwan, M.; Ghebeh, H. PD-L1 promotes OCT4 and Nanog expression in breast cancer stem cells by sustaining PI3K/AKT pathway activation. *Int. J. Cancer* **2017**, *141*, 1402–1412. [CrossRef] [PubMed]

64. Wei, F.; Zhang, T.; Deng, S.-C.; Wei, J.-C.; Yang, P.; Wang, Q.; Chen, Z.-P.; Li, W.-L.; Chen, H.-C.; Hu, H.; et al. PD-L1 promotes colorectal cancer stem cell expansion by activating HMGA1-dependent signaling pathways. *Cancer Lett.* **2019**, *450*, 1–13. [CrossRef] [PubMed]

65. Beatty, G.L.; Gladney, W.L. Immune escape mechanisms as a guide for cancer immunotherapy. *Clin. Cancer Res.* **2014**, *21*, 687–692. [CrossRef] [PubMed]

66. Zhang, B.; Dang, J.; Ba, D.; Wang, C.; Han, J.; Zheng, F. Potential function of CTLA-4 in the tumourigenic capacity of melanoma stem cells. *Oncol. Lett.* **2018**, *16*, 6163–6170. [CrossRef]

67. Mantovani, A.; Dinarello, C.A.; Molgora, M.; Garlanda, C. Interleukin-1 and Related Cytokines in the Regulation of Inflammation and Immunity. *Immunity* **2019**, *50*, 778–795. [CrossRef]

68. Caucheteux, S.M.; Hu-Li, J.; Guo, L.; Bhattacharyya, N.; Crank, M.; Collins, M.T.; Paul, W.E. IL-1b enhances inflammatory TH2 differentiation. *J. Allergy Clin. Immunol.* **2016**, *138*, 898–901. [CrossRef]

69. Li, Y.; Wang, L.; Pappan, L.; Galliher-Beckley, A.; Shi, J. IL-1b promotes stemness and invasiveness of colon cancer cells through Zeb1 activation. *Mol. Cancer* **2012**, *11*, 87. [CrossRef]

70. Siddiqui, I.; Erreni, M.; Kamal, M.A.; Porta, C.; Marchesi, F.; Pesce, S.; Pasqualini, F.; Schiarea, S.; Chiabrando, C.; Mantovani, A.; et al. Differential role of Interleukin-1 and Interleukin-6 in K-Ras-driven pancreatic carcinoma undergoing mesenchymal transition. *OncoImmunology* **2017**, *7*, e1388485. [CrossRef]

71. Fisher, D.T.; Appenheimer, M.M.; Evans, S.S. The two faces of IL-6 in the tumor microenvironment. *Semin. Immunol.* **2014**, *26*, 38–47. [CrossRef]

72. Wei, H. Interleukin 6 signaling maintains the stem-like properties of bladder cancer stem cells. *Transl. Cancer Res.* **2019**, *8*, 557–566. [CrossRef]

73. Wan, S.; Zhao, E.; Kryczek, I.; Vatan, L.; Sadovskaya, A.; Ludema, G.; Simeone, D.M.; Zou, W.; Welling, T.H. Tumor-associated macrophages produce interleukin 6 and signal via STAT3 to promote expansion of human hepatocellular carcinoma stem cells. *Gastroenterology* **2014**, *147*, 1393–1404. [CrossRef] [PubMed]

74. Ishibashi, K.; Koguchi, T.; Matsuoka, K.; Onagi, A.; Tanji, R.; Takinami-Honda, R.; Hoshi, S.; Onoda, M.; Kurimura, Y.; Hata, J.; et al. Interleukin-6 induces drug resistance in renal cell carcinoma. *Fukushima J. Med. Sci.* **2018**, *64*, 103–110. [CrossRef] [PubMed]

75. Wang, Y.; Zong, X.; Mitra, S.; Mitra, A.K.; Matei, D.; Nephew, K.P. IL-6 mediates platinum-induced enrichment of ovarian cancer stem cells. *JCI Insight* **2018**, *3*, 122360. [CrossRef] [PubMed]

76. David, J.M.; Dominguez, C.; Hamilton, D.H.; Palena, C. The IL-8/IL-8R axis: A double agent in tumor immune resistance. *Vaccines* **2016**, *4*, 22. [CrossRef]

77. Jin, F.; Miao, Y.; Xu, P.; Qiu, X. IL-8 regulates the stemness properties of cancer stem cells in the small-cell lung cancer cell line H446. *OncoTargets Ther.* **2018**, *11*, 5723. [CrossRef] [PubMed]

78. Chen, L.; Fan, J.; Chen, H.; Meng, Z.; Chen, Z.; Wang, P.; Liu, L. The IL-8/CXCR1 axis is associated with cancer stem cell-like properties and correlates with clinical prognosis in human pancreatic cancer cases. *Sci. Rep.* **2014**, *4*, 5911. [CrossRef]

79. Singh, J.K.; Simões, B.M.; Howell, S.J.; Farnie, G.; Clarke, R. Recent advances reveal IL-8 signaling as a potential key to targeting breast cancer stem cells. *Breast Cancer Res.* **2013**, *15*, 210. [CrossRef]

80. Salomon, B.L.; Leclerc, M.; Tosello, J.; Ronin, E.; Piaggio, E.; Cohen, J.L. Tumor Necrosis Factor +¦ and Regulatory T Cells in Oncoimmunology. *Front. Immunol.* **2018**, *9*, 444. [CrossRef]

81. Wang, H.; Wang, H.S.; Zhou, B.H.; Li, C.L.; Zhang, F.; Wang, X.F.; Zhang, G.; Bu, X.Z.; Cai, S.H.; Du, J. Epithelial-mesenchymal transition (EMT) induced by TNF-+¦ requires AKT/GSK-3+¦-mediated stabilization of snail in colorectal cancer. *PLoS ONE* **2013**, *8*, e56664. [CrossRef]

82. Jinushi, M.; Sato, M.; Kanamoto, A.; Itoh, A.; Nagai, S.; Koyasu, S.; Dranoff, G.; Tahara, H. Milk fat globule epidermal growth factor–8 blockade triggers tumor destruction through coordinated cell-autonomous and immune-mediated mechanisms. *J. Exp. Med.* **2009**, *206*, 1317–1326. [CrossRef] [PubMed]

83. Jinushi, M.; Chiba, S.; Yoshiyama, H.; Masutomi, K.; Kinoshita, I.; Dosaka-Akita, H.; Yagita, H.; Takaoka, A.; Tahara, H. Tumor-associated macrophages regulate tumorigenicity and anticancer drug responses of cancer stem/initiating cells. *Proc. Natl. Acad. Sci. USA* **2011**, *108*, 12425–12430. [CrossRef] [PubMed]

84. Wolffe, A.P.; Matzke, M.A. Epigenetics: Regulation Through Repression. *Science* **1999**, *286*, 481–486. [CrossRef] [PubMed]

85. Luger, K.; Mäder, A.W.; Richmond, R.K.; Sargent, D.F.; Richmond, T.J. Crystal structure of the nucleosome core particle at 2.8 Å resolution. *Nature* **1997**, *389*, 251–260. [CrossRef]

86. Jenuwein, T.; Allis, C.D. Translating the histone code. *Science* **2001**, *293*, 1074–1080. [CrossRef]

87. McCulloch, E.A.; Siminovitch, L.; Till, J.; Locke, M. Spleen-Colony Formation in Anemic Mice of Genotype WW. *Science* **1964**, *144*, 844–846. [CrossRef]

88. Seita, J.; Weissman, I.L. Hematopoietic stem cell: Self-renewal versus differentiation. *Wiley Interdiscip. Rev. Syst. Boil. Med.* **2010**, *2*, 640–653. [CrossRef]

89. Schwartzentruber, J.; Korshunov, A.; Liu, X.-Y.; Jones, D.T.W.; Pfaff, E.; Jacob, K.; Sturm, D.; Fontebasso, A.M.; Quang, D.A.K.; Tönjes, M.; et al. Driver mutations in histone H3.3 and chromatin remodelling genes in paediatric glioblastoma. *Nature* **2012**, *482*, 226–231. [CrossRef]

90. Wu, C.; Allis, C.D. Nucleosomes, histones & chromatin part B. Preface. *Methods Enzymol.* **2012**, *513*, xv–xvi.

91. Funato, K.; Major, T.; Lewis, P.; Allis, C.D.; Tabar, V. Use of human embryonic stem cells to model pediatric gliomas with H3.3K27M histone mutation. *Science* **2014**, *346*, 1529–1533. [CrossRef]

92. Band, V.; Zhao, X.; Malhotra, G.K.; Band, H. Shared signaling pathways in normal and breast cancer stem cells. *J. Carcinog.* **2011**, *10*, 38. [CrossRef]

93. Handle, F.; Erb, H.H.H.; Luef, B.; Hoefer, J.; Dietrich, D.; Parson, W.; Kristiansen, G.; Santer, F.; Culig, Z. SOCS3 Modulates the Response to Enzalutamide and is Regulated by AR Signaling and CpG Methylation in Prostate Cancer Cells. *Mol. Cancer Res.* **2016**, *14*, 574–585. [CrossRef] [PubMed]

94. Andersson, E.R.; Sandberg, R.; Lendahl, U. Notch signaling: Simplicity in design, versatility in function. *Development* **2011**, *138*, 3593–3612. [CrossRef] [PubMed]

95. Ghoshal, P.; Nganga, A.J.; Moran-Giuati, J.; Szafranek, A.; Johnson, T.R.; Bigelow, A.J.; Houde, C.M.; Avet-Loiseau, H.; Smiraglia, D.J.; Ersing, N.; et al. Loss of the SMRT/NCoR2 Corepressor Correlates with JAG2 Overexpression in Multiple Myeloma. *Cancer Res.* **2009**, *69*, 4380–4387. [CrossRef] [PubMed]

96. Gregorieff, A.; Clevers, H. Wnt signaling in the intestinal epithelium: From endoderm to cancer. *Genes Dev.* **2005**, *19*, 877–890. [CrossRef] [PubMed]

97. Jones, P.A.; Baylin, S.B. The fundamental role of epigenetic events in cancer. *Nat. Rev. Genet.* **2002**, *3*, 415–428. [CrossRef]

98. Ying, J.; Li, H.; Yu, J.; Ng, K.M.; Poon, F.F.; Wong, S.C.; Chan, A.T.; Sung, J.J.; Tao, Q. WNT5A exhibits tumor-suppressive activity through antagonizing the Wnt/beta-catenin signaling, and is frequently methylated in colorectal cancer. *Clin. Cancer Res.* **2008**, *14*, 55–61. [CrossRef]

99. Laird, P.W. Cancer epigenetics. *Hum. Mol. Genet.* **2005**, *14*, R65–R76. [CrossRef]

100. Ansel, K.M.; Lee, D.U.; Rao, A. An epigenetic view of helper T cell differentiation. *Nat. Immunol.* **2003**, *4*, 616–623. [CrossRef]

101. Bergman, Y.; Cedar, H. A stepwise epigenetic process controls immunoglobulin allelic exclusion. *Nat. Rev. Immunol.* **2004**, *4*, 753–761. [CrossRef]

102. Smale, S.T.; Fisher, A.G. Chromatin structure and gene regulation in the immune system. *Annu Rev Immunol.* **2002**, *20*, 427–462. [CrossRef] [PubMed]

103. Sigalotti, L.; Fratta, E.; Coral, S.; Maio, M. Epigenetic drugs as immunomodulators for combination therapies in solid tumors. *Pharmacol. Ther.* **2014**, *142*, 339–350. [CrossRef] [PubMed]

104. Héninger, E.; Krueger, T.E.; Lang, J.M. Augmenting Antitumor Immune Responses with Epigenetic Modifying Agents. *Front. Immunol.* **2015**, *6*, 29. [CrossRef] [PubMed]

105. Terranova-Barberio, M.; Thomas, S.; Munster, P.N. Epigenetic modifiers in immunotherapy: A focus on checkpoint inhibitors. *Immunotherapy* **2016**, *8*, 705–719. [CrossRef] [PubMed]

106. Chang, C.C.; Campoli, M.; Restifo, N.P.; Wang, X.; Ferrone, S. Immune selection of hot-spot beta 2-microglobulin gene mutations, HLA-A2 allospecificity loss, and antigen-processing machinery component down-regulation in melanoma cells derived from recurrent metastases following immunotherapy. *J. Immunol.* **2005**, *174*, 1462–1471. [CrossRef] [PubMed]

107. Fratta, E.; Coral, S.; Covre, A.; Parisi, G.; Colizzi, F.; Danielli, R.; Nicolay, H.; Sigalotti, L.; Maio, M. The biology of cancer testis antigens: Putative function, regulation and therapeutic potential. *Mol. Oncol.* **2011**, *5*, 164–182. [CrossRef] [PubMed]

108. Wrangle, J.; Wang, W.; Koch, A.; Easwaran, H.; Mohammad, H.P.; Vendetti, F.; VanCriekinge, W.; Demeyer, T.; Du, Z.; Parsana, P.; et al. Alterations of immune response of non-small cell lung cancer with Azacytidine. *Oncotarget* **2013**, *4*, 2067–2079. [CrossRef]

109. Booth, L.; Roberts, J.L.; Sander, C.; Lee, J.; Kirkwood, J.M.; Poklepovic, A.; Dent, P. The HDAC inhibitor AR42 interacts with pazopanib to kill trametinib/dabrafenib-resistant melanoma cells in vitro and in vivo. *Oncotarget* **2017**, *8*, 16367–16386. [CrossRef]

110. Beg, A.A.; E Gray, J. HDAC inhibitors with PD-1 blockade: A promising strategy for treatment of multiple cancer types? *Epigenomics* **2016**, *8*, 1015–1017. [CrossRef]

111. Yang, H.; Lan, P.; Hou, Z.; Guan, Y.; Zhang, J.; Xu, W.; Tian, Z.; Zhang, C. Histone deacetylase inhibitor SAHA epigenetically regulates miR-17-92 cluster and MCM7 to upregulate MICA expression in hepatoma. *Br. J. Cancer* **2014**, *112*, 112–121. [CrossRef]

112. Baylin, S.B.; Jones, P.A. A decade of exploring the cancer epigenome — Biological and translational implications. *Nat. Rev. Cancer* **2011**, *11*, 726–734. [CrossRef] [PubMed]

113. Ahuja, N.; Easwaran, H.; Baylin, S.B. Harnessing the potential of epigenetic therapy to target solid tumors. *J. Clin. Investig.* **2014**, *124*, 56–63. [CrossRef] [PubMed]

114. Chiappinelli, K.B.; Strissel, P.L.; Desrichard, A.; Li, H.; Henke, C.; Akman, B.; Hein, A.; Rote, N.S.; Cope, L.M.; Snyder, A.; et al. Inhibiting DNA Methylation Causes an Interferon Response in Cancer via dsRNA Including Endogenous Retroviruses. *Cell* **2015**, *162*, 974–986. [CrossRef] [PubMed]

115. Krummel, M.F.; Allison, J.P. CD28 and CTLA-4 have opposing effects on the response of T cells to stimulation. *J. Exp. Med.* **1995**, *182*, 459–465. [CrossRef]

116. Topalian, S.L.; Hodi, F.S.; Brahmer, J.R.; Gettinger, S.N.; Smith, D.; McDermott, D.F.; Powderly, J.D.; Carvajal, R.D.; Sosman, J.A.; Atkins, M.B.; et al. Safety, activity, and immune correlates of anti-PD-1 antibody in cancer. *New Engl. J. Med.* **2012**, *366*, 2443–2454. [CrossRef]
117. Collins, D.C.; Sundar, R.; Lim, J.S.; Yap, T.A. Towards Precision Medicine in the Clinic: From Biomarker Discovery to Novel Therapeutics. *Trends Pharmacol. Sci.* **2017**, *38*, 25–40. [CrossRef] [PubMed]
118. Hasin, Y.; Seldin, M.; Lusis, A.J. Multi-omics approaches to disease. *Genome Boil.* **2017**, *18*, 83. [CrossRef] [PubMed]

MDPI
St. Alban-Anlage 66
4052 Basel
Switzerland
Tel. +41 61 683 77 34
Fax +41 61 302 89 18
www.mdpi.com

Cells Editorial Office
E-mail: cells@mdpi.com
www.mdpi.com/journal/cells

www.ingramcontent.com/pod-product-compliance
Lightning Source LLC
LaVergne TN
LVHW070628170726
843515LV00004B/205